# Conscious Life

John S. Torday • Rob G. Sacco

# Conscious Life

## Bridging Cellular Origins, Quantum Realities, and Cosmic Connections

Springer

John S. Torday
David Geffen School of Medicine
University of California, Los Angeles
Los Angeles, CA, USA

Rob G. Sacco
Fibonacci LifeChart
Toronto, ON, Canada

ISBN 978-3-031-80499-1 ISBN 978-3-031-80500-4 (eBook)
https://doi.org/10.1007/978-3-031-80500-4



This Springer imprint is published by the registered company Springer Nature Switzerland AG
The registered company address is: Gewerbestrasse 11, 6330 Cham, Switzerland

*To the seekers of knowledge,*
*who look beyond the seen to understand*
*the unseen.*
*To the pioneers of science,*
*who trace the origins of life from the smallest*
*cell to the vast cosmos.*
*To those who believe that consciousness is*
*not merely a product of biology,*
*but a bridge between matter and meaning.*
*May this work inspire a deeper connection*
*between science and spirit,*
*between evolution and awareness,*
*and between the individual and the infinite.*

—John S. Torday & Rob G. Sacco

# Prelude

*When we do math, or physics, or chemistry,*
*or on the other hand, we write poetry*
*or a musical ditty,*
*we're acknowledging we're parts of the 'whole'.*
*And the 'whole' is greater than the sum of its parts.*
*All the mathing, physicsing, chemistrying,*
*All Arts, painting, and sculpting, musicing and poesing,*
*'Couldn't put Humpty together again'*
*Why so?*
*What's the problem?*
*Lack of 'hobgoblins' of our soul*
*Probably so.*
*So what to do?*
*Pray for inspiration*
*From above?*
*Find the answer in love?*
*I say look inward for inspiration.*
*Our physiology's got the answer.*
*Every cell inscribed with conscience,*
*aspires, perspires and sweats out the Truth.*
*—John S. Torday*

# Preface

The journey of science often begins with the pursuit of understanding life's most profound questions: How did life emerge? What is consciousness? How are we, as biological beings, connected to the Cosmos? As scientists, our paths have led us to explore these questions from complementary perspectives, converging on a shared vision that unites cellular biology, evolutionary theory, Quantum Mechanics, and Cosmology into an integrative framework for understanding life and consciousness.

Our collaboration began as a conversation between two minds deeply invested in unraveling the mysteries of existence. One of us, John Torday, brought decades of research in cellular physiology and evolutionary biology, uncovering the profound role of symbiogenesis and cellular communication in shaping life. The other, Rob Sacco, approached these questions through the lens of complex systems, mathematical modeling, and the universal patterns that connect the microcosm of cells to the macrocosm of the Universe.

"Conscious Life" is the culmination of that dialogue, a synthesis of ideas that transcends disciplinary boundaries. In this book, we explore the origins of life as a cellular phenomenon and trace its evolution into consciousness, shaped by the interplay of biological structures, Quantum Entanglement, and Cosmic principles. We delve into the mathematical underpinnings of life, from the elegance of Fibonacci sequences in cellular organization, to Gödel's incompleteness theorems as a metaphor for the open-ended nature of consciousness. At its core, this book reflects our shared belief that the story of life is one of unity of systems, processes, and meaning.

This work is not merely an academic exercise, but an invitation to a larger audience to see life as a continuum, a bridge between the physical and metaphysical. Whether you are a scientist, philosopher, or curious reader, our aim is to offer insights into how the same principles governing the Cosmos manifest in our own biology. Understanding these principles is not just about intellectual fulfillment; it is about reconnecting with the profound sense of belonging to a greater whole.

The journey we outline in this book mirrors the process of life itself—a dynamic interplay of parts forming an emergent whole. It is a journey that has humbled and inspired us, reinforcing the belief that the answers we seek are as much about discovering the unity in complexity as they are about challenging our perspectives.

We hope "Conscious Life" serves as a bridge for you, just as our collaboration has been for us—a connection between disciplines, ideas, and ultimately, between you and the Cosmos.

With gratitude for the systems that make inquiry possible,

Los Angeles, CA, USA John S. Torday
Toronto, ON, Canada Rob G. Sacco

# About the Book

"Conscious Life" offers a groundbreaking exploration of the intricate tapestry that connects the cellular origins of life, quantum dynamics, and the cosmos. At its heart, the book investigates consciousness as a phenomenon rooted in cellular processes, shaped by quantum mechanics, and expressed through evolutionary and cosmic patterns. From the first micelles that sparked life, to the geometric and mathematical principles underlying cellular structures, it bridges biology, physics, and philosophy in an integrative narrative.

Drawing on the latest research in molecular biology, quantum physics, and evolutionary theory, the book examines how processes like symbiogenesis, quantum entanglement, and mathematical frameworks such as Gödel's theorems and the Fibonacci sequence underpin both life and the Universe. It proposes a vision of evolution as a symphony of interconnected systems, where the local meets the nonlocal, and the finite merges with the infinite.

Through an accessible yet profound synthesis of science and philosophy, "Conscious Life" reveals how understanding these connections offers insights not only into the nature of but also into our place within an ever-expanding Cosmos. This book is essential reading for those seeking to uncover the universal principles that bind matter, mind, and meaning.

# Contents

## Part II Epigenetics: The Mechanism of Evolution

# Part I
# Energy and Evolution: The Basis of Life

# Chapter 1
# The Cellular Birth of Consciousness

The journey to understand consciousness has often focused on the human brain, overlooking its evolutionary roots. Consciousness did not emerge fully formed; rather, it evolved over billions of years through incremental adaptations at the cellular level. By examining the origin of life, the transition to multicellular organisms, and the molecular processes underlying homeostasis and adaptation, we can trace the developmental trajectory of consciousness. This chapter begins with the birth of cellular consciousness, exploring how life adapted to environmental challenges and how these adaptations laid the groundwork for higher-order awareness.

## The Evolutionary Roots of Consciousness

The origins of consciousness lie deep within the evolutionary history of life. Long before the development of complex brains or nervous systems, the earliest forms of life exhibited behaviors and mechanisms that laid the foundation for what we now recognize as awareness. This section examines the role of micelles as precursors to cellular life, the evolutionary pressures that drove the development of calcium regulation, and the connection between cellular homeostasis and emergent consciousness.

### *Micelles: The First Step Toward Life*

Micelles, simple lipid-based structures, represent a critical precursor to cellular life. These amphiphilic molecules, with hydrophilic and hydrophobic poles, self-assemble in aqueous environments to form spherical structures (Deamer, 2017).



J. S. Torday, R. G. Sacco, *Conscious Life*,
https://doi.org/10.1007/978-3-031-80500-4_1

This natural self-organization created compartments that shielded internal chemical reactions from external conditions, providing an environment conducive to the emergence of life.

The ability of micelles to maintain stability in fluctuating environments marks the earliest form of homeostasis. By isolating chemical processes, micelles introduced a rudimentary boundary between "self" and "other," an essential characteristic of living systems. This primitive organization foreshadowed the development of cellular membranes and the mechanisms required for more complex forms of survival.

## *Calcium Regulation and Cellular Awareness*

As life began to emerge in Earth's ancient oceans, environmental challenges shaped its evolution. Among these challenges, the high concentrations of calcium ions ($Ca^{2+}$) posed a significant threat due to their reactivity and toxicity. Early cells evolved regulatory mechanisms to mitigate this toxicity, including calcium pumps and buffering systems (Case et al., 2007). These mechanisms represent an early form of environmental sensing and response—a foundational aspect of cellular awareness.

Calcium regulation not only protected cells from ion-induced damage but also enabled the development of intracellular signaling pathways. These pathways are critical for coordinating cellular functions, and would later become integral to neural signaling in multicellular organisms. By responding dynamically to calcium levels, cells demonstrated an early capacity for adaptive behavior, a precursor to higher forms of consciousness.

## *Homeostasis as a Foundation for Consciousness*

At the heart of cellular survival lies homeostasis, the ability to maintain internal stability despite external changes. This capacity is critical for life, and represents a key link between cellular mechanisms and the emergence of consciousness. Storr et al. (2013) argue that the regulatory processes underlying homeostasis reflect a rudimentary form of awareness, enabling cells to sense and adapt to environmental conditions.

For example, single-celled organisms like amoebae exhibit behaviors such as chemotaxis, moving toward nutrients, or away from toxins. These actions are driven by biochemical pathways that detect environmental signals and orchestrate appropriate responses. Though simple, these behaviors illustrate the evolutionary roots of consciousness as an adaptive strategy for survival.

### *Conclusion*

The evolutionary roots of consciousness are deeply embedded in the molecular and cellular mechanisms that enabled life to adapt and thrive. From the compartmentalization of micelles to the regulatory systems that manage calcium toxicity and maintain these early adaptations set the stage for the emergence of awareness. Consciousness, at its core, is an extension of life's fundamental drive to survive and interact dynamically with its environment.

## Symbiogenesis and the Birth of Multicellular Consciousness

The transition from unicellular to multicellular life marked a transformative leap in the evolution of consciousness. This shift enabled the development of more complex behaviors and systems, including cell-cell communication and cooperative survival strategies, which laid the foundation for multicellular awareness. By exploring the role of cell-cell communication, the process of Symbiogenesis, and the dual temporal dimensions of consciousness—synchronic and diachronic—this section traces how multicellular life advanced the evolutionary trajectory of awareness.

### *Cell–Cell Communication as a Competitive Adaptation*

Unicellular organisms, such as bacteria, often exhibit behaviors that mimic multicellularity. Prokaryotic cells use mechanisms like quorum sensing and biofilm formation to coordinate activities in response to environmental stimuli. Quorum sensing involves the release and detection of signaling molecules that regulate group behaviors, such as resource acquisition or defense against threats. Similarly, biofilms—structured communities of bacteria—exemplify cooperative living arrangements that enhance survival in challenging environments (Torday & Rehan, 2012).

The emergence of true cell–cell communication in multicellular organisms likely evolved as a competitive response to these prokaryotic strategies. Unlike the temporary cooperation seen in biofilms, multicellular organisms rely on permanent and highly integrated communication systems. These systems, mediated by chemical signals, electrical impulses, and structural interactions, enable the synchronization of cellular functions, laying the groundwork for higher-order processes. This ability to communicate and coordinate in real time is essential for multicellular consciousness.

## *Symbiogenesis: Transforming Threats into Advantages*

The process of symbiogenesis was pivotal in the evolutionary leap from unicellular to multicellular life. Symbiogenesis involves the fusion of distinct organisms into a single, cooperative entity. This process not only allowed cells to survive environmental threats but also transformed these threats into physiological advantages. For instance, the endosymbiotic incorporation of mitochondria into ancestral eukaryotic cells provided a reliable energy source, enabling the metabolic demands of multicellularity to be met (Sagan, 1967).

Symbiogenesis also established the cellular basis of "first-order" consciousness, which refers to the ability of an organism to respond adaptively to its environment. By integrating foreign elements into their physiology, cells demonstrated a capacity for innovation and resilience. These traits are foundational to the emergent properties of multicellular life, where individual cells work collectively to maintain the homeostasis and survival of the organism (Torday, 2015).

## *Synchronic and Diachronic Dimensions of Consciousness*

Multicellularity introduced both synchronic (real-time) and diachronic (historical) dimensions of consciousness. Synchronic consciousness allows organisms to respond dynamically to immediate environmental stimuli, while diachronic consciousness integrates historical information, enabling long-term adaptations. These two dimensions are exemplified in embryogenesis, where the development of an organism reflects its evolutionary history.

During embryonic development, cells communicate and differentiate in a coordinated manner, guided by genetic and epigenetic instructions. This process mirrors the evolutionary transitions that shaped multicellular life, providing a living model of how synchronic and diachronic processes converge. The interplay between real-time signaling and historical programming underscores the complexity of multicellular consciousness (Torday & Rehan, 2009).

## *Conclusion*

The evolution of multicellular life represents a significant advancement in the development of consciousness. Through cell–cell communication, symbiogenesis, and the integration of synchronic and diachronic processes, multicellularity created new possibilities for awareness and adaptation. By transforming environmental threats into opportunities and coordinating activities across multiple dimensions, multicellular organisms set the stage for the emergence of higher-order consciousness.

## Molecular Evidence for Consciousness

The molecular and genetic underpinnings of life offer profound insights into the origins and nature of consciousness. At the heart of this exploration are the mechanisms that govern embryonic development and their alignment with evolutionary processes. By examining the role of Hox genes, the relationship between genetic expression and evolutionary history, and the interplay between genetic mechanisms and homeostasis, we uncover the molecular foundation of consciousness.

### *Hox Genes and the Blueprint of Development*

Hox genes, a subset of homeobox genes, are critical in embryonic development. These transcription factors orchestrate the spatial and temporal expression of genes that shape the body axis of an organism. Notably, Hox genes exhibit colinearity: their arrangement on chromosomes reflects their order of expression along the anterior-posterior axis of the developing organism. This precise alignment ensures that tissues and organs form in the correct sequence and location (Torday & Rehan, 2012).

The colinearity of Hox genes underscores the deep connection between molecular signals and the physical structure of the body. This molecular choreography is not only vital for development but also exemplifies how genetic instructions translate into physiological complexity. In the context of consciousness, this process illustrates how molecular signals form the structural and functional basis of awareness.

### *Developmental Processes and Evolutionary History*

The mechanisms that guide development are also a window into evolutionary history. Hox genes and other genetic pathways reflect the accumulated adaptations of countless generations. By regulating the formation of body structures, these genes encode the evolutionary transitions that led to increasingly complex organisms. Torday and Rehan (2012) argue that this genetic continuity provides a framework for understanding consciousness as a continuum from molecular to cognitive levels.

For example, the regulatory networks established by Hox genes are conserved across diverse species, highlighting their evolutionary significance. This conservation indicates that the same genetic mechanisms underpin fundamental aspects of life, from simple organisms to humans. By aligning developmental processes with evolutionary history, these genetic systems create a seamless link between past and present, enabling the emergence of higher-order consciousness.

### *Genetic Expression, Homeostasis, and Awareness*

The relationship between genetic expression and homeostasis is central to understanding how molecular processes contribute to consciousness. Homeostasis involves maintaining internal stability in response to external changes, a process mediated by precise genetic regulation. When genetic expression is disrupted, the resulting imbalances can manifest as "dis-ease" or altered states of awareness (Storr et al., 2013).

For instance, oxidative stress and DNA damage disrupt cellular signaling pathways, impairing the body's ability to adapt to environmental challenges. These disruptions not only affect physiological health but also influence mental states, emphasizing the interplay between genetic mechanisms and awareness. The ability of genetic systems to maintain homeostasis is thus integral to sustaining both physical and cognitive functions.

### *Conclusion*

The molecular and genetic markers of consciousness highlight the intricate relationship between development, evolution, and awareness. Through the precise regulation of Hox genes, the alignment of genetic mechanisms with evolutionary history, and the maintenance of homeostasis, life establishes the foundation for consciousness. These molecular processes, rooted in the origins of life, illustrate how awareness emerges as a continuum from the cellular to the cognitive, bridging the physical and abstract dimensions of existence.

## Consciousness as Healing and Adaptation

Consciousness is not limited to the cognitive processes of complex organisms, but extends to the cellular and systemic levels, manifesting in adaptive and reparative mechanisms. This "second-order" consciousness is evident in processes like healing, cellular adaptation, and phenomena that transcend immediate physical presence, such as phantom limb sensations. These mechanisms underscore the integrative nature of life, where biological systems interact dynamically with their environment to maintain equilibrium and respond to challenges.

### *Healing and Adaptation as Cellular Consciousness*

Healing and adaptation are fundamental processes that highlight the capacity of biological systems to maintain homeostasis and respond to damage. At the cellular level, healing involves orchestrated responses to injury, including the activation of signaling pathways, mobilization of resources, and restoration of function. Torday and Rehan (2009) propose that such reparative mechanisms reflect a form of cellular awareness, as cells "perceive" damage and initiate processes to restore equilibrium.

Adaptation further demonstrates this cellular awareness. For example, cells exposed to hypoxic conditions adapt by upregulating genes that enhance oxygen transport and utilization. This dynamic response to environmental stressors underscores the ability of life to not only survive but also optimize function in changing conditions. These adaptive processes, which occur at the cellular and systemic levels, are integral to second-order consciousness, where the organism's awareness extends beyond immediate threats to include long-term survival strategies.

## *Oxidative Stress and DNA Repair: Cellular Awareness in Action*

Oxidative stress, caused by an imbalance between reactive oxygen species (ROS) and the body's ability to detoxify them, represents a significant challenge to cellular integrity. When oxidative stress occurs, it can damage cellular components, including lipids, proteins, and DNA. Cells respond to this threat by activating repair mechanisms that mitigate damage and restore function. These responses illustrate a form of cellular "awareness," as cells detect disruptions and act to resolve them (Storr et al., 2013).

DNA repair mechanisms, in particular, highlight the sophistication of cellular adaptation. When DNA damage is detected, cells halt the cell cycle to prevent the propagation of errors, activate repair enzymes, and restore genomic integrity. This process not only protects the organism from mutations and disease but also reflects an intrinsic capacity for self-regulation and adaptation. The ability to sense, evaluate, and respond to damage underscores the continuity between cellular processes and the broader phenomenon of consciousness.

## *Phantom Limb Phenomena: Evidence of Persistent Awareness*

Phantom limb phenomena provide compelling evidence of consciousness persisting beyond immediate physical presence. These phenomena, experienced by individuals who have lost limbs, involve sensations such as pain, itching, or movement in the absent limb. Torday and Rehan (2012) argue that such experiences are maintained through upstream signaling pathways in the brain and nervous system, which retain a "map" of the body even in the absence of physical structures.

This persistence of awareness demonstrates the integrative nature of consciousness, where the organism's perception extends beyond the physical state to include memory, signaling, and adaptation. Phantom limb phenomena underscore the role of systemic consciousness in maintaining a cohesive sense of self, even when physical continuity is disrupted. This capacity reflects the evolutionary advantage of second-order consciousness, enabling organisms to adapt and respond dynamically to their environment and internal states.

### *Conclusion*

Second-order consciousness, as manifested in healing, adaptation, and phenomena like phantom limb sensations, highlights the complexity and integrative nature of awareness. These processes illustrate how biological systems interact with their environment and maintain equilibrium through adaptive and reparative mechanisms. From cellular responses to oxidative stress, and DNA damage, to the persistence of awareness in phantom limb phenomena, second-order consciousness underscores the dynamic interplay between biological systems and their environment. By extending our understanding of consciousness beyond cognition, we gain deeper insights into the evolutionary and systemic foundations of awareness.

## The Evolutionary Synthesis of Consciousness

The synthesis of cellular processes and cosmological principles provides a profound framework for understanding consciousness. Life, at its core, is a dynamic interplay of energy, adaptation, and organization, reflecting universal principles that govern both biological and cosmic systems. This section explores the connections between cellular communication, homeostasis, and cosmological dynamics, positioning consciousness as an emergent property rooted in these relationships. It concludes by discussing future directions for an integrative understanding of consciousness that bridges cellular biology, evolutionary theory, and cosmology.

### *Cellular Systems and Cosmological Principles*

Cellular communication systems and homeostatic mechanisms mirror larger cosmological principles such as energy regulation and entropy. At the cellular level, communication occurs through signaling molecules, ion channels, and structural interactions, ensuring that organisms maintain internal stability and respond to environmental changes. These processes exemplify the concept of energy conservation—resources are dynamically allocated to optimize survival and maintain balance.

Similarly, in cosmology, energy flows drive the organization of matter and the emergence of complex systems. From the formation of stars to the structure of galaxies, the universe operates under principles of energy regulation and entropy reduction. Torday (2015) argues that the same principles govern biological systems, suggesting that life's evolution is deeply interconnected with the physics of the cosmos.

For example, homeostasis—the cellular capacity to maintain equilibrium—can be likened to the universe's tendency toward thermodynamic equilibrium. While the universe as a whole moves toward increasing entropy, localized systems, such as galaxies or living organisms, temporarily reverse this trend through processes that create order. This alignment between cellular and cosmic dynamics reveals a fundamental continuity between life and the universe.

## *Recursive Relationships: Cells and the Cosmos*

The relationship between cellular and universal dynamics is inherently recursive. Life adapts to its environment by incorporating external challenges into internal processes, creating a feedback loop that fosters complexity and awareness. This recursive relationship positions consciousness as an emergent property of life's interactions with its surroundings (Torday, 2015).

For instance, cellular evolution can be viewed as a microcosm of cosmic evolution. Just as the universe began with simple elements that coalesced into complex structures, life started with simple cells that evolved into multicellular organisms. This process of increasing complexity is driven by the interplay between energy constraints, environmental pressures, and the capacity for adaptation.

Embryogenesis provides a striking example of this recursive relationship. During development, cells replicate and differentiate according to genetic instructions, reflecting the organism's evolutionary history. This process parallels the self-organizing dynamics of the universe, where patterns emerge from the interaction of fundamental forces. Such parallels reinforce the idea that consciousness, as a product of cellular processes, is deeply connected to the fabric of the cosmos.

## *Consciousness as an Emergent Property*

Consciousness, in this framework, is not a static trait but an emergent property arising from the dynamic interplay of biological and cosmic forces. It is the result of systems interacting to optimize energy use, adapt to environmental challenges, and maintain order amidst entropy. Torday and Rehan (2012) emphasize that the homeostatic mechanisms underpinning life are themselves reflective of the universe's organizing principles.

The emergent nature of consciousness can be seen in its adaptive flexibility. From the regulation of calcium ions in early cells to the intricate signaling networks of the human brain, consciousness evolves alongside the systems it governs. This adaptability is rooted in the same principles that guide the formation of stars, planets, and galaxies, highlighting the universality of these processes.

## *Future Directions: Integrative Approaches*

Understanding consciousness requires integrative approaches that connect cellular biology, evolutionary theory, and cosmology. By exploring the parallels between life's mechanisms and universal dynamics, researchers can develop holistic models that account for the emergent nature of awareness.

One promising direction is the study of Quantum Biology, which investigates how quantum phenomena influence biological processes. Torday (2015) suggests that the principles of Quantum Mechanics, such as entanglement and superposition, may provide insights into the recursive relationships that underpin consciousness.

Additionally, advances in Systems Biology and computational modeling offer tools for simulating the complex interactions between cellular and cosmic systems. These approaches can help elucidate how consciousness emerges from the integration of molecular, cellular, and environmental dynamics.

Finally, interdisciplinary research bridging biology, physics, and philosophy can deepen our understanding of the connections between life and the universe. Such efforts have the potential to not only explain the origins of consciousness but also shed light on its role in the broader context of existence.

## *Conclusion*

The evolutionary synthesis of consciousness reveals the profound interconnectedness of life and the cosmos. Cellular communication and homeostatic mechanisms reflect the same principles of energy regulation and entropy reduction that govern the universe. Through recursive relationships, life incorporates environmental challenges into its processes, fostering the emergence of complexity and awareness. Consciousness, as an emergent property, is a testament to the dynamic interplay between biological and cosmic systems. By embracing integrative approaches, we can move closer to understanding the origins and nature of consciousness, bridging the gap between the microcosm of cells and the macrocosm of the universe.

# References


Case, R. M., Eisner, D., Gurney, A., Jones, O., Muallem, S., & Verkhratsky, A. (2007). Evolution of calcium homeostasis: From birth of the first cell to an omnipresent signaling system. *Cell Calcium, 42*(4–5), 345–350.

Deamer, D. (2017). The role of lipid membranes in life's origin. *Life, 7*(1), 5.

Sagan, L. (1967). On the origin of mitosing cells. *Journal of Theoretical Biology, 14*(3), 255–274.

Storr, S. J., Woolston, C. M., Zhang, Y., & Martin, S. G. (2013). Redox environment, free radical, and oxidative DNA damage. *Antioxidants & Redox Signaling, 18*(18), 2399–2408.

Torday, J. S. (2015). Homeostasis as the mechanism of evolution. *Biology, 4*(3), 573–590.

Torday, J. S., & Rehan, V. K. (2009). Lung evolution as a cipher for physiology. *Physiological Genomics, 38*(1), 1–6.

Torday, J. S., & Rehan, V. K. (2012). *Evolutionary biology, cell-cell communication, and complex disease*. Wiley.

# Chapter 2
# Micelles: The Seed of Conscious Life

The origins of life—and, by extension, consciousness—can be traced back to the most fundamental building blocks of biology. Among these are micelles, simple lipid-based structures that played a pivotal role in the transition from abiotic chemistry to the first living cells. Often overlooked in discussions of life's beginnings, micelles represent a remarkable example of self-organization, encapsulating the foundational principles that underpin all living systems.

This chapter explores the significance of micelles as the precursors to cellular life, examining their ability to compartmentalize and protect biochemical reactions. It delves into the chemical properties of amphiphilic molecules that enable micelle formation, their role in fostering the emergence of homeostasis, and the evolutionary pressures that made micelles indispensable for the development of life. By understanding micelles as the seed of conscious life, we gain insight into how the basic principles of self-organization, stability, and environmental interaction laid the groundwork for the emergence of cellular awareness and, ultimately, consciousness itself.

From their spontaneous assembly in Earth's primordial oceans to their role as the forerunners of cell membranes, micelles exemplify the remarkable interplay between structure and function. This chapter demonstrates that the study of micelles not only illuminates the origins of life but also provides a crucial foundation for understanding the evolutionary trajectory of consciousness. By exploring this often-overlooked phase of life's history, we uncover the profound connections between the simplest forms of biological organization and the complex phenomena of awareness and adaptation that characterize living systems today.



J. S. Torday, R. G. Sacco, *Conscious Life*,
https://doi.org/10.1007/978-3-031-80500-4_2

## The Origins of Life and the Role of Micelles

The emergence of life on Earth represents one of the most profound transitions in the history of the cosmos. Central to this transformation is the micelle, a lipid-based structure that served as the foundation for the first living systems. Micelles provided the critical architecture necessary for compartmentalization, the separation of internal and external environments, and the organization of biochemical reactions. This section examines the physicochemical properties of lipids that enable micelle formation, their connection to universal principles such as gravitational forces and energy dynamics, and their role as the interface between David Bohm's explicate and implicate orders, offering a unified perspective on the origins of life.

### *Physicochemical Properties of Lipids and Micelle Formation*

Lipids, amphiphilic molecules with hydrophilic (water-attracting) heads and hydrophobic (water-repelling) tails, are key to understanding micelle formation. In aqueous environments, lipids spontaneously organize into micelles, with their hydrophobic tails sequestered inside and their hydrophilic heads exposed to the surrounding water. This self-assembly process, driven by the minimization of free energy, creates a semipermeable boundary that can encapsulate biochemical reactions (Deamer, 2017).

Micelles provided the earliest proto-cells with the ability to compartmentalize, protecting fragile prebiotic molecules from the chaotic external environment. This compartmentalization was essential for maintaining stability, enabling chemical reactions to occur in a controlled microenvironment. By acting as the first semipermeable membranes, micelles set the stage for the evolution of more complex cellular structures, including bilayer membranes, which later defined true cellular life (Claassen & Spooner, 1996).

### *Micelle Formation and Gravitational Forces*

The formation of micelles is not only a biochemical phenomenon but also one influenced by fundamental physical forces. Einstein's Field Theory, which describes the curvature of spacetime due to gravitational forces, provides an intriguing lens through which to view micelle formation. The curved surfaces of micelles are akin to the warped geometry of spacetime, where energy gradients enable processes that counteract entropy (Hawking, 1998).

This analogy extends to the energy dynamics of micelles. The curved lipid surfaces create localized energy gradients, allowing the system to harness and organize energy in a way that maintains negative entropy—a hallmark of living systems. By stabilizing chemical reactions within their boundaries, micelles provided the energetic foundation necessary for life's emergence, bridging the physical laws of the cosmos with the nascent properties of biology.

### *Micelles as the Interface Between Explicate and Implicate Orders*

David Bohm's concept of explicate and implicate orders offers a philosophical framework for understanding micelles as the initial interface of life. The explicate order represents the manifest world of form and structure, while the implicate order describes the underlying, unbroken wholeness of reality (Bohm, 1980). Micelles, as emergent structures from simple lipid molecules, embody this duality. They arise from the implicate order of molecular properties and self-organize into explicate forms that interact dynamically with their environment.

Micelles thus serve as a bridge between the potentiality of the implicate order and the manifested complexity of life. Their ability to compartmentalize and maintain stability underpins the transition from nonliving matter to living systems. By establishing the initial conditions for life's emergence, micelles illustrate the seamless integration of universal principles with biological evolution.

### *Conclusion*

The micelle represents the foundational structure upon which the earliest forms of life were built. Through their unique physicochemical properties, micelles provided the first semipermeable membranes, enabling compartmentalization and energy management critical for life. Their curved surfaces, influenced by gravitational forces, created energy gradients that counteracted entropy, aligning them with fundamental cosmological principles. Moreover, as an interface between Bohm's explicate and implicate orders, micelles encapsulate the profound interplay between potentiality and manifestation that defines the origins of life. These simple structures laid the groundwork for the emergence of cellular systems and the evolutionary trajectory toward consciousness.

## Symbiogenesis and Quantum Entanglement in Cellular Evolution

The micelle, as one of the earliest building blocks of life, represents a critical intersection between environmental adaptation and Quantum Dynamics. Its ability to self-organize, adapt, and maintain stability paved the way for the evolution of more complex systems. This section explores how symbiogenesis enabled micelles to assimilate environmental threats, how quantum entanglement fostered interdependence and self-organization within protocells, and how the principles of physics, such as Newton's Third Law, supported the stabilization of early cellular structures.

## *Symbiogenesis: Environmental Threats as Evolutionary Catalysts*

Symbiogenesis, the fusion of distinct biological entities to create new cooperative systems, played a transformative role in early evolution. Initially, micelles encountered various environmental threats, such as fluctuating temperatures, chemical gradients, and ionic imbalances. Rather than succumbing to these challenges, micelles harnessed them as opportunities for adaptation and innovation.

Lynn Sagan (1967) described symbiogenesis as a process by which simpler structures merged to form more complex ones, incorporating environmental elements to enhance survival. In the context of micelles, this process facilitated the integration of foreign molecules, such as ions and amino acids, into their structure. By assimilating these external factors, micelles transformed potential threats into functional components, driving evolutionary progress.

Symbiogenesis provided micelles with the capacity for innovation, setting the stage for protocellular development. These evolutionary leaps were not random; they were shaped by interactions with the environment, demonstrating the profound adaptability of early life forms. This adaptability underscores the micelle's role as a precursor to more advanced cellular structures.

## *Quantum Entanglement and Protocellular Self-Organization*

Quantum entanglement offers a compelling framework for understanding the internal dynamics of micelles and protocells. Entanglement occurs when particles become interconnected, such that the state of one particle instantaneously influences the state of another, regardless of distance. Torday (2023a) proposes that quantum entanglement played a pivotal role in fostering interdependence among the components of protocells, enabling the emergence of self-organization.

Within micelles, entangled particles, such as ions and water molecules, created a network of interactions that stabilized the structure and facilitated coordinated behavior. These interactions allowed micelles to maintain internal coherence while adapting to external fluctuations. Quantum entanglement, therefore, was not merely a physical phenomenon, but a driver of biological complexity, enabling the micelle to function as a cohesive unit.

By promoting interdependence among its components, Entanglement enhanced the micelle's ability to self-organize, a property fundamental to all living systems. This capacity for self-organization marked a critical transition from inanimate matter to life, bridging the Quantum world with biological evolution.

## *Micelles as Early Models of Homeostasis*

The micelle's ability to maintain internal stability in the face of environmental changes established it as the first example of homeostasis. Homeostasis, the dynamic process of regulating internal conditions to sustain life, is a defining feature of all living organisms. For micelles, this meant maintaining structural integrity and functionality amidst external perturbations.

Torday (2015) links the principles of homeostasis to Newton's Third Law of Motion, which states that for every action, there is an equal and opposite reaction. In the context of micelles, this principle is evident in their ability to counteract external forces with internal adjustments. For example, when exposed to ionic imbalances, micelles can rearrange their molecular components to stabilize their structure. This dynamic interplay between internal and external forces exemplifies the foundational principles of homeostasis.

Micelles demonstrated an early form of feedback regulation, where changes in their environment triggered compensatory responses. These responses not only stabilized the micelle but also set the stage for the evolution of more sophisticated homeostatic mechanisms in later life forms.

## *Conclusion*

The micelle's role in cellular evolution is a testament to its remarkable adaptability and complexity. Through symbiogenesis, micelles assimilated environmental threats into functional components, driving evolutionary innovation. Quantum entanglement fostered interdependence and self-organization within these structures, bridging the quantum and biological realms. Finally, the micelle's capacity for homeostasis, grounded in Newtonian physics, provided the stability necessary for the emergence of life. Together, these processes illustrate the micelle's critical role as a foundational structure in the evolutionary trajectory toward cellular complexity and consciousness.

## The Micelle's Role in Cosmic and Biological Integration

The micelle is not only a foundational structure in the emergence of cellular life but also serves as a bridge between biological systems and the cosmos. Its organization and function reflect universal natural laws, linking the microcosmic processes of life to the macrocosmic principles governing the universe. This section explores how the arrangement of elements within micelles mirrors cosmic processes like stellar nucleosynthesis, how the micelle's compliance with universal laws underpins the emergence of consciousness, and how its properties prefigure nonlocal connectivity, exemplifying the unity of life with the cosmos.

### *Micelles and Cosmic Order: Parallels with Stellar Nucleosynthesis*

The organizational principles that govern the formation of micelles echo those found in the cosmos. In stellar nucleosynthesis, the elements essential for life—carbon, oxygen, hydrogen, and nitrogen—are forged in the cores of stars. These elements are then distributed across the universe through supernova explosions, creating the raw materials for planetary and biological systems (Torday & Rehan, 2017).

Similarly, micelles organize these elements into functional structures, assembling lipid molecules into a spherical form that encapsulates and protects biochemical reactions. This self-organization reflects a continuity between the physical processes of the cosmos and the biological processes of life. Torday and Rehan (2017) suggest that this alignment underscores a fundamental principle: the same forces that shape the universe also drive the evolution of life. The micelle, therefore, is not an isolated phenomenon, but a manifestation of cosmic order at the biological level.

### *Universal Laws and the Foundation of Consciousness*

The micelle's formation and function are governed by universal natural laws, such as thermodynamics and energy minimization, which apply across scales from the atomic to the cosmic. These laws enable the micelle to maintain stability and facilitate the emergence of complexity, laying the groundwork for consciousness. Einstein and Born (2004) emphasized the universality of these laws, describing them as the framework within which all physical and biological phenomena occur.

By adhering to these universal principles, the micelle established the foundation for cellular processes that would later evolve into awareness. Torday (2023b) argues that consciousness arises from the alignment of individual systems, such as cells, with these larger natural laws. The micelle's compliance with these principles not only enabled its survival but also connected it to the broader dynamics of the universe, positioning it as the precursor to individual and cosmic consciousness.

This connection suggests that consciousness is not a uniquely biological trait, but an emergent property of the universe itself. The micelle's ability to organize energy, maintain homeostasis, and adapt to environmental changes reflects a fundamental unity between life and the cosmos, where the principles guiding the formation of galaxies also shape the evolution of awareness.

## *Nonlocality and Universal Connectivity*

The micelle's properties also prefigure the concept of nonlocality, a phenomenon central to Quantum Mechanics. Nonlocality, famously described by Einstein as "spooky action at a distance," refers to the instantaneous connection between particles regardless of distance. This idea extends to biological systems, where connectivity and communication occur beyond immediate physical interactions (Hawking, 1998).

In the context of micelles, nonlocality can be seen in their ability to integrate external environmental information into internal processes. For example, the self-organization of lipid molecules within a micelle is influenced by external forces such as temperature, ionic concentration, and chemical gradients. This dynamic interaction reflects the micelle's role as a conduit for universal connectivity, bridging the internal and external worlds.

Torday (2023b) further posits that the micelle's ability to maintain coherence amidst environmental fluctuations mirrors the nonlocality observed in Quantum Systems. This coherence forms the basis for cellular communication, which later evolved into complex signaling networks in multicellular organisms. By acting as an early model of nonlocal connectivity, the micelle exemplifies the interplay between biological and cosmic systems, highlighting the interconnectedness of life and the universe.

## *Conclusion*

The micelle serves as a bridge between life and the cosmos, reflecting the organizational principles that govern both biological and cosmic systems. Its formation and function mirror stellar nucleosynthesis, demonstrating a profound continuity between the elements of life and the universe. By adhering to universal natural laws, the micelle established the foundation for consciousness, connecting individual awareness to cosmic dynamics. Additionally, its properties prefigure nonlocality, emphasizing its role in universal connectivity. As a foundational structure, the micelle embodies the unity of life with the cosmos, offering a model for understanding the emergence of consciousness as an intrinsic property of the universe.

## The Micelle as a Model for Evolutionary Mechanisms

The micelle, as one of the earliest self-organizing biological structures, offers a powerful framework for understanding evolutionary mechanisms. It demonstrates how complexity arises in response to environmental challenges, serves as a metaphorical link to mathematical models like knot theory, and provides insight into evolutionary innovations such as magnetoreception. This section explores how the micelle embodies key principles of evolutionary adaptation, and sheds light on the processes that drove the emergence of life.

### *From Micelles to Unicellular Life: Increasing Complexity*

The transition from micelles to unicellular organisms represents a process of increasing complexity driven by environmental pressures. Micelles provided a protected environment for chemical reactions, enabling the development of early biochemical pathways. As these structures encountered dynamic environmental challenges—such as fluctuating temperatures, nutrient availability, and chemical gradients—they evolved mechanisms to maintain homeostatic stability while adapting to their surroundings.

Gould and Vrba (1982) introduced the concept of exaptation, which describes how structures originally evolved for one function can be co-opted for another. Micelles exemplify this principle. Initially formed through the self-assembly of lipids, micelles transitioned from passive compartments to active systems capable of responding to their environment. For example, the incorporation of molecules such as amino acids and nucleotides into micelles created protocells, the precursors to unicellular organisms. These adaptations highlight the micelle's role as a stepping stone toward the complexity required for cellular life.

### *Knot Theory and Cellular Adaptation*

The process of micelle formation can be linked to mathematical models such as knot theory, which examines the properties of loops and their transformations. Kauffman (2006) proposed that the untangling and reconfiguring of knots provides a useful analogy for understanding biological adaptation. In this framework, micelles can be seen as "knots" that self-organize into stable forms while retaining the flexibility to adapt to environmental changes.

Knot theory offers insights into the dynamic nature of micelles. Just as knots can be tightened, loosened, or restructured, micelles undergo constant reorganization in response to external forces. For example, changes in temperature or ionic concentration can cause lipid molecules to rearrange, altering the structure and function of the micelle. This adaptability mirrors the broader evolutionary processes that drive the emergence of complexity.

The mathematical elegance of knot theory underscores the interconnectedness of biological and physical systems. By linking micelle formation to these abstract models, we gain a deeper understanding of how simple structures evolve into complex systems capable of responding dynamically to their environment.

## *Magnetoreception and Evolutionary Innovation*

Magnetoreception, the ability of organisms to sense magnetic fields, provides a striking example of micelle-mediated evolutionary innovation. This phenomenon is mediated by cryptochrome, a light-sensitive protein found in many organisms, including birds and insects. Cryptochrome interacts with magnetic fields through quantum processes, enabling organisms to navigate their environment (McFadden & Khalili, 2016).

The micelle's role in this evolutionary innovation lies in its capacity to compartmentalize and stabilize the chemical reactions necessary for magnetoreception. Early micelles provided the protected environment needed for the evolution of cryptochrome and other complex biochemical systems. By enabling the development of proteins with magnetic sensitivity, micelles facilitated the emergence of a biological navigation system that connected molecular processes to environmental cues.

Magnetoreception illustrates how micelles contributed to evolutionary adaptations that enhanced survival. By serving as a platform for the development of molecular machinery, micelles linked the biochemistry of life to broader environmental dynamics. This interplay between structure, function, and environment exemplifies the principles of evolutionary innovation.

## *Conclusion*

The micelle serves as a model for understanding evolutionary mechanisms, illustrating how complexity arises through self-organization and environmental adaptation. The transition from micelles to unicellular organisms highlights the role of exaptation in driving innovation, while knot theory provides a metaphorical framework for understanding the flexibility and adaptability of micelles. Additionally, the micelle's contribution to magnetoreception demonstrates its role in linking molecular processes to environmental navigation. By examining the micelle through these lenses, we gain valuable insights into the processes that shaped the evolution of life and laid the groundwork for biological complexity.

## The Micelle's Legacy: From Niche Construction to Modern Civilization

The micelle, as a foundational structure in the emergence of life, left an indelible mark on the evolutionary trajectory of organisms and the development of human societies. By acting as the first example of niche construction, micelles established a dynamic interplay between life and the environment, creating a blueprint for how organisms shape and are shaped by their surroundings. This legacy extends beyond cellular evolution, offering valuable insights into human behavior, societal organization, and humanity's relationship with nature. In this section, we examine how the principles underlying micelle evolution mirror human development, resource management, and the balance between nature and civilization, emphasizing their relevance in addressing global challenges like climate change.

### *Micelles as the First Niche Construction*

Niche construction refers to the process by which organisms modify their environment to improve their chances of survival. The micelle represents the earliest form of niche construction, creating a controlled microenvironment that shielded internal biochemical processes from external chaos (Torday, 2016). By compartmentalizing and stabilizing prebiotic reactions, micelles established a continuum between the environment and life, where internal processes were dynamically influenced by external forces.

This relationship illustrates a fundamental principle of life: the ability to actively shape one's environment in response to challenges. For the micelle, this meant forming semipermeable boundaries that selectively interacted with the external world. This interaction was not static; micelles evolved to incorporate environmental elements, such as ions and molecules, into their internal structure, creating an adaptive feedback loop. This capacity for environmental modification is echoed in more complex organisms, including humans, who continue to alter their surroundings to enhance survival and productivity.

### *From Cellular Evolution to Human Behavior*

The transition from micelles to unicellular and multicellular life mirrors patterns observed in human behavior and societal development. Just as micelles organized resources to maintain stability and adapt to environmental pressures, human societies have developed intricate systems for managing resources, from agriculture to energy production. Torday and Miller (2017) suggest that the principles of cellular evolution, such as cooperation, resource optimization, and niche construction, are reflected in human societal structures.

For example, the cooperative behavior of cells in a multicellular organism parallels the interdependence of individuals within a society. Both systems rely on efficient communication, resource allocation, and adaptation to external challenges. However, the ability to modify the environment on a global scale introduces new complexities. While micelles operated within the constraints of Natural Laws, human societies often disrupt ecological balances, leading to challenges such as resource depletion and environmental degradation.

This dynamic underscores the need to align human activity with the principles of balance and sustainability observed in nature. By understanding the micelle's legacy, we can derive lessons for mitigating the environmental impacts of modern civilization.

## *The Micelle's Enduring Significance*

The micelle's legacy extends beyond its role in cellular evolution to encompass broader questions about consciousness, evolution, and humanity's place in nature. As the first structure to bridge the gap between life and the environment, the micelle provides a model for understanding the interconnectedness of all living systems. This interconnectedness is critical for addressing global challenges, such as climate change, which require a holistic perspective that integrates biology, ecology, and human activity.

Torday and Miller (2017) emphasize that the principles of niche construction, homeostasis, and adaptation exemplified by the micelle can guide efforts to restore ecological balance. For instance, strategies to combat climate change, such as reforestation and renewable energy adoption, reflect the micelle's ability to adaptively modify its environment while maintaining stability. By applying these principles, humanity can align its activities with the natural processes that sustain life.

Furthermore, the micelle's significance extends to understanding consciousness as an emergent property of life's interaction with its environment. The principles of self-organization and adaptation that define the micelle are foundational to the development of awareness, both at the cellular level and in complex organisms. This perspective positions the micelle as a unifying concept for exploring the origins of life, evolution, and the balance between humanity and nature.

## *Conclusion*

The micelle's legacy as the first niche constructor highlights its enduring significance in the trajectory of life and civilization. From creating a continuum between life and the environment to mirroring human societal development, the micelle provides a powerful framework for understanding the principles of adaptation, cooperation, and balance. By reflecting on the micelle's role in evolutionary history, we gain valuable insights into addressing modern challenges, such as climate change, and fostering a sustainable relationship between humanity and nature. As we confront the complexities of the Anthropocene, the micelle reminds us of the profound interconnectedness of life and the need to align human activity with the principles that sustain the biosphere.

## References

Bohm, D. (1980). *Wholeness and the implicate order*. Routledge.

Claassen, E., & Spooner, D. (1996). Lipid membranes in the origin of life. *Origins of Life and Evolution of Biospheres, 26*(1), 61–79.

Deamer, D. (2017). The role of lipid membranes in life's origin. *Life, 7*(1), 5.

Einstein, A., & Born, M. (2004). *The Born-Einstein letters: Correspondence between Albert Einstein and Max and Hedwig Born from 1916 to 1955*. Macmillan.

Gould, S. J., & Vrba, E. S. (1982). Exaptation—A missing term in the science of form. *Paleobiology, 8*(1), 4–15.

Hawking, S. W. (1998). *A brief history of time*. Bantam Books.

Kauffman, S. (2006). *Reinventing the sacred: A new view of science, reason, and religion*. Basic Books.

McFadden, J., & Khalili, F. (2016). Quantum biology and the microscopic magnetic compass. *Physics Today, 69*(2), 46–51.

Sagan, L. (1967). On the origin of mitosing cells. *Journal of Theoretical Biology, 14*(3), 255–274.

Torday, J. S. (2015). Homeostasis as the mechanism of evolution. *Biology (Basel), 4*, 573–590.

Torday, J. S. (2016). The cell as the first niche construction. *Biology (Basel), 5*, 19.

Torday, J. S. (2023a). Consciousness, embodied quantum entanglement. *Progress in Biophysics and Molecular Biology, 177*, 125–128.

Torday, J. S. (2023b). The holism of cosmology and consciousness. *Progress in Biophysics and Molecular Biology, 177*, 181–184.

Torday, J. S., & Miller, W. B., Jr. (2017). The resolution of ambiguity as the basis for life. *Progress in Biophysics and Molecular Biology, 131*, 288–297.

Torday, J. S., & Rehan, V. K. (2017). *Evolution, the logic of biology*. Wiley.

# Chapter 3
# Evolution's Blueprint: The Principles of Cellular Change

At the heart of evolution lies a simple yet profound truth: change is the cornerstone of life. This chapter explores the cellular mechanisms that underpin the adaptability and resilience of living systems, offering insights into how evolution operates at the most fundamental level. From the first protocells to the intricate multicellular organisms of today, the principles governing cellular change have not only shaped life's diversity but also reveal the shared blueprint underlying all biological systems.

The evolution of life is a story of problem-solving. Cells, the basic units of life, respond to environmental challenges with remarkable ingenuity, employing mechanisms such as gene regulation, signaling pathways, and metabolic adaptation. These processes, which began with single-celled organisms billions of years ago, illustrate the capacity of life to innovate, survive, and thrive.

This chapter delves into the principles of cellular change that drive evolution, including adaptation, exaptation, and niche construction. It examines how cells co-opt existing structures for new purposes, create environments that enhance their survival, and integrate internal and external signals to drive evolutionary progress. These cellular strategies not only highlight the ingenuity of evolution but also provide a blueprint for understanding the continuity between molecular processes and the emergence of complex organisms.

By tracing the cellular principles of evolution, this chapter reveals how the dynamics of change operate across all scales of life, from single cells to entire ecosystems. It positions cellular mechanisms as the architects of life's evolutionary journey, providing a foundation for understanding the interconnectedness and adaptability that define the living world.



J. S. Torday, R. G. Sacco, *Conscious Life*,
https://doi.org/10.1007/978-3-031-80500-4_3

## Cellular Memory and the Origins of Evolution

The single cell is not merely a biochemical machine; it embodies a form of consciousness and retains an evolutionary memory that preserves life's history. This memory underpins evolution, enabling cells to adapt, innovate, and respond to existential threats. By exploring experimental evidence for cellular memory, the role of symbiogenesis in integrating challenges, and the philosophical implications of life's monistic origins, this section reveals the profound interconnectedness of all living systems.

### *Experimental Evidence of Cellular Memory*

The concept of cellular memory challenges traditional views of the cell as a reactive entity, positioning it instead as an adaptive and learning system. Cellular memory refers to the ability of cells to retain and utilize information about past experiences to inform future behavior. Experimental studies provide compelling evidence for this phenomenon.

For instance, phenotypic reversibility observed under microgravity conditions demonstrates cellular memory at work. Cells exposed to microgravity revert to ancestral phenotypes, as seen in alterations to gene expression and cellular behavior that mimic early evolutionary states (Torday & Rehan, 2012). This phenomenon suggests that cells retain a latent memory of their evolutionary history, allowing them to reaccess prior states when faced with novel challenges.

Similarly, studies on biofilms, where bacteria exhibit collective memory through shared signaling and coordinated behavior, underscore the role of cellular memory in adaptation. Purevdorj-Gage et al. (2006) showed that yeast can adaptively reorganize in response to environmental stress, demonstrating a form of collective cellular memory that enhances survival.

### *Symbiogenesis and Evolutionary Integration*

Cellular memory also plays a crucial role in integrating existential threats through symbiogenesis, the process by which distinct organisms merge to form new evolutionary entities. Symbiogenesis enables cells to incorporate external elements, such as genes or organelles, into their systems, transforming challenges into opportunities for innovation.

Kutschera (2009) highlights symbiogenesis as a key mechanism in the evolution of complex life. For example, the incorporation of mitochondria into ancestral eukaryotic cells not only enhanced energy production but also established a lasting partnership that became foundational to multicellular life. This evolutionary leap was facilitated by the cell's ability to integrate and retain the memory of this symbiotic relationship, ensuring its perpetuation across generations.

Through cellular memory, symbiogenesis becomes more than a reactive process; it is a forward-looking strategy that preserves the lessons of past challenges. This capacity for integration reflects the cell's intrinsic awareness and adaptability, positioning it as a conscious agent in the evolutionary narrative.

## *Philosophical Implications of Life's Monistic Origins*

The origins of life, as reflected in cellular memory, align with the philosophical concept of monism—the idea that all existence is interconnected and arises from a single underlying reality. Loy (1997) suggests that life's monistic nature resolves the philosophical question of interconnectedness, offering a scientific basis for the sense of "something greater" that many perceive.

Cellular memory exemplifies this interconnectedness. By retaining the evolutionary history of life, cells act as repositories of shared experience, linking past, present, and future. This continuity underscores the unity of all living systems, where individual cells contribute to the collective trajectory of evolution.

The philosophical implications of cellular memory extend to our understanding of consciousness. If life's origins are monistic, then the awareness embodied by a single cell is not separate from the broader phenomenon of consciousness. Instead, it represents an emergent property of life's intrinsic interconnectedness, where cellular memory serves as both a bridge to the past and a foundation for future evolution.

## *Conclusion*

Cellular memory is a cornerstone of evolutionary biology, enabling cells to retain their history, integrate challenges, and contribute to the continuity of life. Experimental evidence, such as phenotypic reversibility and biofilm behavior illustrates the cell's ability to adapt and innovate. Symbiogenesis further highlights how cellular memory transforms existential threats into opportunities for evolutionary advancement. Finally, the monistic origins of life, reflected in cellular memory, emphasize the interconnectedness of all living systems and the profound unity of life and consciousness. This foundational concept sets the stage for understanding the cell as both a historical repository, and an active agent in the evolution of complexity.

## The Role of Gravity in Cellular Evolution

Gravitational forces, an omnipresent feature of the Earth's environment, have played a critical role in shaping the structure and evolution of life from its earliest stages. Gravity influenced the alignment of lipids during the formation of micelles, the precursor to cellular structures, and later impacted the transition to multicellular organisms and complex physiological systems. This section explores the role of gravity in cellular evolution, from its impact on micelle formation to its influence on multicellularity, bipedalism, and physiological homeostasis.

### *Gravity and Micelle Formation: The Birth of Protocells*

The first step in the evolution of cellular life was the formation of micelles, lipid-based structures that provided a protective boundary for biochemical reactions. Gravitational forces played a significant role in aligning lipids and promoting the self-organization of these structures in early aquatic environments (Deamer, 2017). As lipids aggregated in water, their amphiphilic properties caused them to align with their hydrophobic poles inward and hydrophilic poles outward, forming stable, spherical micelles bounded by semipermeable membranes.

Gravity influenced this process by facilitating the sedimentation and spatial organization of lipid molecules, creating localized environments conducive to micelle formation. These gravitational effects were critical for concentrating lipids and other biomolecules in shallow water environments, increasing the likelihood of micelle formation and the eventual emergence of protocells. By providing a physical framework for the alignment and stabilization of these early structures, gravity laid the groundwork for cellular evolution.

### *Gravity and the Transition to Multicellularity*

As life evolved from single cells to multicellular organisms, gravity continued to shape biological structures and processes. In multicellularity, cells cooperate and communicate to form tissues and organs that can withstand gravitational forces. Gravity influenced the organization of these structures by imposing physical constraints that selected for efficient and adaptive designs.

One example of gravity's influence is the development of body plans that optimize structural integrity and energy use in different gravitational contexts. In aquatic environments, where buoyancy offsets gravitational effects, early multicellular organisms exhibited simpler, more symmetrical forms. As life transitioned to terrestrial environments, gravity imposed new challenges, leading to adaptations such as skeletal systems and more complex tissue organization.

Bipedalism in humans provides a later example of gravity-driven evolutionary innovation. Gould and Vrba (1982) describe bipedalism as an exaptation, a trait initially evolved for one function (locomotion) that became co-opted for another (energy-efficient movement under gravitational constraints). By standing upright, humans minimized energy expenditure during movement, freeing resources for other physiological and cognitive functions.

### *Gravity, Cellular Topology, and Physiological Homeostasis*

Gravity's influence extends beyond structural organization to cellular topology and physiological homeostasis. Cellular structures, such as cytoskeletal networks, evolved to counteract gravitational forces, maintaining the shape and stability of cells. The cytoskeleton acts as an internal scaffold that resists deformation, allowing cells to maintain their integrity under gravitational stress (Torday, 2003).

At the physiological level, gravity affects fluid dynamics, nutrient transport, and metabolic processes. For example, gravity influences blood circulation and the distribution of nutrients and oxygen in multicellular organisms. Physiological systems evolved to optimize these processes, maintaining homeostasis despite the challenges posed by gravity. This dynamic interplay between gravity and biological systems highlights its central role in shaping the evolution of life.

### *Conclusion*

Gravity has been a guiding force in the evolution of cellular structures and processes. From facilitating lipid alignment and micelle formation in early protocells, to shaping multicellular body plans and physiological systems, gravity's influence is deeply embedded in the history of life. Its impact on cellular topology, multicellularity, and adaptations like bipedalism underscores its role as both a physical constraint and a driver of evolutionary innovation. By understanding the role of gravity in cellular evolution, we gain insights into the interconnectedness of physics and biology, revealing the fundamental forces that have shaped life on Earth.

## Symbiogenesis and Quantum Entanglement

The evolution of life has been driven by processes that integrate environmental challenges into biological systems, promoting adaptation and innovation. Symbiogenesis, the incorporation of foreign entities into cellular structures, and quantum entanglement, the interconnectedness of particles, represent key mechanisms in cellular evolution. Together, they highlight the dynamic interplay between biology and the fundamental principles of the cosmos. This section explores how symbiogenesis and quantum entanglement have shaped cellular evolution, linking these processes to the cosmological principles underlying the fabric of existence.

## *Symbiogenesis: Assimilation and Innovation*

Symbiogenesis is a transformative evolutionary mechanism that allows cells to assimilate environmental threats into their physiology, turning challenges into opportunities for innovation. Lynn Sagan (1967) first articulated symbiogenesis as the process by which distinct organisms merged to form new biological entities. A classic example is the incorporation of bacteria to form mitochondria in early eukaryotic cells, which enabled more efficient energy production and laid the foundation for complex multicellular life.

In this context, micelles, as precursors to cellular structures, represent an early form of symbiotic integration. These lipid-based compartments provided a safe environment for biochemical reactions, shielding molecules from external threats while incorporating useful resources from the environment. Over time, micelles evolved mechanisms to integrate external elements, such as ions and amino acids, into their internal processes, reflecting the principles of symbiogenesis (Torday, 2022).

Symbiogenesis extends beyond molecular interactions to encompass entire ecosystems. It exemplifies how life thrives by integrating external factors into its systems, promoting evolutionary leaps that enhance survival and complexity. This capacity for innovation underscores the role of symbiogenesis as a cornerstone of cellular evolution.

## *Quantum Entanglement and the Leap to Homeostasis*

Quantum entanglement, a phenomenon where particles become interconnected such that the state of one influences the state of another instantaneously, offers a compelling framework for understanding the dynamics of micelle formation and cellular evolution. Torday (2023) argues that the self-organization observed in micelle formation parallels the interconnected behavior of entangled particles, emphasizing their homologous roles in achieving homeostasis.

Micelles, formed through the self-assembly of lipid molecules, demonstrate a form of Quantum Coherence, where their components work in unison to maintain stability. This coherence enables micelles to adapt to external fluctuations while preserving their structural integrity. The principles of quantum entanglement may underlie this coordinated behavior, suggesting that cellular systems harness Quantum phenomena to achieve balance and functionality.

The integration of Quantum processes into cellular evolution bridges the microcosmic and macrocosmic realms, highlighting the unity of life and the universe. By leveraging quantum entanglement, early cellular structures like micelles demonstrated the capacity for self-regulation, setting the stage for the emergence of more complex homeostatic mechanisms in multicellular organisms.

## *Cosmological Connections: Micelles and Space-Time*

The processes that drive cellular evolution are not isolated from the principles governing the universe. Einstein's concept of gravity as a distortion of space-time provides an intriguing parallel to micelle formation. Just as gravity warps space-time to create localized regions of attraction and organization, micelles generate microenvironments that encapsulate and stabilize biochemical reactions.

Torday (2023) posits that micelle formation reflects the fundamental dynamics of the cosmos, where matter and energy interact to create order amidst chaos. This comparison underscores the continuity between cellular evolution and cosmological principles, suggesting that the forces shaping life are rooted in the same Natural Laws that govern the universe.

By framing micelle formation as a microcosmic reflection of gravitational dynamics, we gain a deeper understanding of the interconnectedness between biology and cosmology. The emergence of life, from micelles to complex organisms, can thus be seen as an extension of the universe's capacity for self-organization and adaptation.

## *Conclusion*

Symbiogenesis and euantum entanglement represent central mechanisms in cellular evolution, illustrating how life integrates external challenges into its systems to achieve innovation and stability. Symbiogenesis enabled cells to assimilate environmental threats, transforming them into evolutionary advantages, while quantum entanglement facilitated the coordinated behavior necessary for homeostasis. Furthermore, the parallels between micelle formation and cosmological principles, such as Einstein's space-time distortion, highlight the unity of biological and universal dynamics. Together, these processes reveal the profound interconnectedness between life and the cosmos, offering a holistic perspective on the origins and evolution of cellular systems.

## Mathematics and the Logic of Evolution

Mathematics, often referred to as the universal language, provides profound insights into the principles driving evolution. Cellular processes, far from being random, adhere to mathematical frameworks that reflect the structure and logic of the cosmos. This section examines the Fibonacci sequence as a metaphor for cellular evolution, explores mathematical homologies like knot theory in understanding cellular behaviors, and proposes that the mathematical architecture of the universe is mirrored in cellular physiology through processes like symbiogenesis.

### *The Fibonacci Sequence: Evolution as a Continuum*

The Fibonacci sequence, a series of numbers where each term is the sum of the two preceding numbers, has long been associated with natural patterns, from the arrangement of leaves to the spirals of galaxies. Tegmark (2014) suggests that the Fibonacci sequence offers a metaphor for cellular evolution, illustrating how the past and present influence the future.

In cellular biology, the Fibonacci sequence can be seen in the iterative and cumulative nature of evolutionary processes. Each step in evolution builds upon previous adaptations, integrating genetic information and environmental challenges. For example, the development of multicellularity required the integration of ancestral traits from unicellular organisms, creating a continuum where each stage informs the next. This interplay of past, present, and future reflects the recursive dynamics inherent in the Fibonacci sequence.

Moreover, the sequence's mathematical elegance mirrors the efficiency of cellular systems. Cells optimize their resources and interactions, much like the Fibonacci sequence optimizes growth and distribution in natural systems. This alignment underscores the idea that evolution operates within a mathematical framework that balances order and adaptability.

### *Mathematical Homologies: Knot Theory and Cellular Behavior*

Knot theory, a branch of mathematics that studies the properties of loops and their configurations, provides a powerful analogy for cellular behaviors during development and reproduction. Kauffman (2006) argues that knot theory reflects the dynamic processes of cellular organization, where structural complexity emerges through the untangling and reconfiguration of molecular networks.

During cell division, for instance, DNA strands must untangle and replicate in a precise manner to ensure accurate genetic inheritance. This process resembles the mathematical principles of knot theory, where loops and intersections are systematically reorganized to achieve stability. Similarly, in cellular development, signaling pathways and molecular interactions form intricate networks that guide differentiation and tissue formation. These networks, like mathematical knots, exhibit flexibility and resilience, adapting to environmental changes while maintaining structural coherence.

The application of knot theory to cellular biology highlights the interconnectedness of mathematics and life, demonstrating how abstract principles can elucidate the complexity of biological systems.

## *The Cosmos in Cells: Symbiogenesis and Mathematical Mirroring*

Rowlands (2015) posits that the mathematical structure of the cosmos is mirrored in the physiology of cells, suggesting that evolution is guided by universal principles. Symbiogenesis, the fusion of distinct organisms to create new evolutionary entities, exemplifies this mirroring. By integrating external elements, cells embody the same mathematical logic that governs the organization of galaxies and the formation of complex systems.

For example, the geometry of lipid bilayers in cellular membranes reflects the principles of minimal energy configurations, a concept rooted in mathematics. Similarly, the fractal-like organization of cellular structures, such as mitochondria, echoes the self-similar patterns observed in Cosmic phenomena. These parallels suggest that the processes driving cellular evolution are not isolated from the universe, but are extensions of its underlying mathematical order.

Symbiogenesis further illustrates this connection by transforming environmental challenges into opportunities for innovation. This capacity for adaptation aligns with the mathematical principles of symmetry breaking and self-organization, which are fundamental to both biology and cosmology. By embracing these principles, cells bridge the gap between microcosmic physiology and macrocosmic dynamics.

## *Conclusion*

Mathematics provides a unifying framework for understanding the logic of evolution, from cellular processes to the structure of the cosmos. The Fibonacci sequence illustrates the recursive nature of evolution, where past and present adaptations shape the future. Knot theory offers insights into cellular behaviors, highlighting the dynamic interplay of structure and function. Finally, the mathematical architecture of the universe is mirrored in cellular physiology, underscoring the continuity between life and the cosmos. By exploring these mathematical frameworks, we gain a deeper appreciation of evolution as a process governed by universal principles that transcend biological boundaries.

## Fractals, Holism, and the Continuum of Evolution

The evolution of life is not merely a series of discrete events but a continuum that spans scales, from molecular interactions to cosmic principles. Fractal patterns in physiology, the "hard problem of consciousness," and the concept of homeostasis as a unifying principle reveal the holistic nature of evolution. This section synthesizes themes from previous discussions—cellular memory, gravity, and symbiogenesis—into a cohesive framework, emphasizing the interconnectedness of life and the universe. It concludes by exploring future directions in understanding evolution as a continuum that integrates cellular biology, physics, and cosmology.

### *Fractals and the Interconnectedness of Evolution*

Fractals, patterns that repeat across different scales, offer a visual and conceptual representation of the interconnectedness inherent in cellular evolution. These self-similar patterns are not confined to mathematics or nature, but extend to physiological systems, illustrating the continuity of evolutionary processes.

Blaisdell et al. (2013) highlight how fractal structures in physiology, such as the branching of blood vessels and the architecture of lungs, reflect underlying principles of energy efficiency and optimal resource distribution. These patterns are rooted in the same fractal dynamics observed in galaxies and river networks, suggesting a universal framework that governs both living and nonliving systems.

In cellular evolution, fractal organization emerges through processes like symbiogenesis and adaptation. For instance, the integration of bacteria to form mitochondria by eukaryotic cells mirrors the fractal logic of nested systems, where smaller components contribute to the function of larger structures. This fractal continuity underscores the unity of life across scales, bridging cellular physiology with the broader dynamics of evolution.

### *The Hard Problem of Consciousness and Homeostasis*

The "hard problem of consciousness," as articulated by Chalmers (1995), questions how subjective experiences (qualia) arise from physical processes. Cellular biology provides a potential answer by positioning homeostasis as the basis for consciousness. Homeostasis, the ability of cells and organisms to maintain internal stability, is a defining feature of life and a potential foundation for the emergence of awareness.

Torday (2022) argues that homeostasis serves as the link between cellular processes and consciousness. By constantly adapting to environmental changes, cells exhibit a form of awareness that underlies their survival. This intrinsic sensitivity to internal and external states may form the precursors of qualia, where subjective experiences emerge from the dynamic balance of cellular systems.

For example, the regulation of calcium ions—a fundamental process in cellular homeostasis—also plays a critical role in neural signaling and brain function. The continuity between these scales suggests that consciousness arises as an emergent property of life's ability to adapt and self-regulate. This perspective positions homeostasis as a bridge between the physical and experiential realms, offering a potential solution to the hard problem of consciousness.

## *The Continuum of Evolution*

Understanding evolution as a continuum requires integrating insights from cellular biology, physics, and cosmology. Cellular memory preserves the evolutionary history of life, gravity shapes biological structures, and symbiogenesis drives innovation through the integration of external elements. These processes are not isolated; they interact dynamically to create the complex systems that define life.

Torday (2022) emphasizes the need for a holistic framework that connects these themes. By viewing evolution as a continuum, we recognize the interconnectedness of all living systems and their alignment with universal principles. This perspective not only deepens our understanding of life but also provides a blueprint for addressing global challenges, such as environmental sustainability and the preservation of biodiversity.

Future research in this area will likely focus on uncovering the mechanisms that link cellular and cosmic processes. Advances in systems biology, quantum biology, and computational modeling hold promise for exploring the emergent properties of life and consciousness. By synthesizing knowledge across disciplines, we can move closer to a unified theory of evolution that encompasses both the microcosmic and macrocosmic scales.

## *Conclusion*

Fractals, homeostasis, and the continuum of evolution reveal the holistic nature of life's trajectory. Fractal patterns in physiology reflect the interconnectedness of cellular evolution across scales, while homeostasis offers a framework for understanding the emergence of consciousness. By integrating cellular biology, physics, and cosmology, we gain a deeper appreciation of evolution as a unified process that bridges the physical and experiential realms. This synthesis not only advances scientific understanding but also inspires a greater sense of connection between humanity and the universe, offering insights for navigating the challenges of the modern world.

## References


Blaisdell, A. P., Pottenger, B. C., & Torday, J. S. (2013). From heartbeats to health recipes: The role of fractal physiology in the ancestral health movement. *Journal of Evolution and Health, 1*(1), 2.

Chalmers, D. (1995). Facing up to the problem of consciousness. *Journal of Consciousness Studies, 2*, 200–219.

Deamer, D. (2017). The role of lipid membranes in life's origin. *Life (Basel), 7*, 5.

Gould, S. J., & Vrba, E. S. (1982). Exaptation: A missing term in the science of form. *Paleobiology, 8*(1), 4–15.

Kauffman, L. (2006). *Formal knot theory*. Dover Publications.

Kutschera, U. (2009). Symbiogenesis, natural selection, and the dynamic earth. *Theory in Biosciences, 128*(3), 191–203.

Loy, D. (1997). *Nonduality: A study in comparative philosophy*. Humanity Books.

Purevdorj-Gage, B., Sheehan, K. B., & Hyman, L. E. (2006). Effects of low-shear modeled micro-gravity on cell function, gene expression, and phenotype in *Saccharomyces cerevisiae*. *Applied and Environmental Microbiology, 72*(7), 4569–4575.

Rowlands, P. (2015). *The foundations of physical law*. World Scientific.

Sagan, L. (1967). On the origin of mitosing cells. *Journal of Theoretical Biology, 14*(3), 255–274.

Tegmark, M. (2014). *Our mathematical universe*. Vintage.

Torday, J. S. (2003). Parathyroid hormone-related protein as a gravisensor in lung and bone cell biology. *Advances in Space Research, 32*(9), 1569–1576.

Torday, J. S. (2022). Evolution, gravity, and the topology of consciousness. *Progress in Biophysics and Molecular Biology, 174*, 50–54.

Torday, J. S. (2023). Consciousness, embodied quantum entanglement. *Progress in Biophysics and Molecular Biology, 177*, 125–128.

Torday, J. S., & Rehan, V. K. (2012). *Evolutionary biology, cell-cell communication, and complex disease*. Wiley.

# Chapter 4
# Energy Pathways: Ontogeny and Phylogeny in Evolution

Evolution is fundamentally a story of energy—its acquisition, distribution, and utilization. From the simplest cells to complex multicellular organisms, the pathways through which life harnesses energy have shaped both individual development (ontogeny) and the broader evolutionary lineage (phylogeny). Energy is not merely a biological necessity; it is the driving force behind adaptation, innovation, and the emergence of complexity.

This chapter explores the intertwined roles of ontogeny and phylogeny in evolutionary processes, focusing on the energy dynamics that underpin life's progression. Ontogeny, the development of an organism from a single cell to maturity, provides a microcosm of evolutionary history, echoing the adaptations that define a species' phylogeny. By examining energy pathways at these scales, we uncover the deep connections between individual development and the evolution of life as a whole.

We begin by tracing the origins of energy acquisition, from the role of photosynthesis and cellular respiration in early life forms to the emergence of specialized systems in multicellular organisms. The chapter then examines how these pathways are conserved and modified across generations, reflecting the balance between stability and innovation that characterizes evolution. Finally, we consider the implications of energy pathways for understanding major evolutionary transitions, from the origin of multicellularity to the development of complex behaviors.

By linking ontogeny and phylogeny through the lens of energy, this chapter provides a framework for understanding the continuity of life, where individual development mirrors and informs the evolutionary trajectory. The energy pathways that sustain life today are not merely remnants of the past, but are active threads weaving together the story of evolution.



J. S. Torday, R. G. Sacco, *Conscious Life*,
https://doi.org/10.1007/978-3-031-80500-4_4

## Cellular Origins and the Singularity of Nature

The origins of life and its subsequent evolution are deeply intertwined with the fundamental forces and materials of the Universe. The hypothesis of the Singularity of Nature posits that life is not an anomaly, but an extension of cosmic processes, where physics and physiology converge to create the conditions necessary for existence. This section explores how environmental forces, stellar nucleosynthesis, and cellular energy flows collectively shaped the emergence of life and established the foundations of evolution.

### *Gravity, Environmental Forces, and Lipid Micelle Formation*

The emergence of life can be traced to the interaction of physical forces and chemical materials in Earth's primordial environment. Gravity, as a universal force, played a significant role in organizing matter, concentrating lipids and other molecules into the conditions necessary for self-organization. Lipid micelles, spherical aggregates of amphiphilic molecules, represent a primordial energy lattice where biochemical reactions could occur in a controlled microenvironment (Torday, 2019).

Environmental forces such as temperature gradients and chemical concentration differences further contributed to micelle formation. These forces created localized pockets of energy, enabling lipids to self-assemble into semipermeable structures capable of compartmentalizing and regulating reactions. Claassen and Spooner (1996) highlighted the unique properties of lipid molecules, whose hydrophobic and hydrophilic poles made them ideal for forming stable membranes. These early micelles served as precursors to protocells, providing the first steps toward organized life.

Gravity's influence extended beyond organization to stabilization. By anchoring molecules in sedimented regions and facilitating their alignment under environmental pressures, gravity acted as a scaffold for the creation of biochemical networks. These networks became the foundation for cellular systems, linking the physical environment to the nascent physiological processes that would drive evolution.

### *Stellar Nucleosynthesis and the Periodic Table*

The chemical elements necessary for life—carbon, hydrogen, calcium, nitrogen, oxygen, phosphorus, sulfur, and iron—were forged in the interiors of stars through stellar nucleosynthesis. When stars reached the end of their life cycles, these elements were dispersed into the cosmos through supernova explosions, seeding the universe with the building blocks of life (Hawking, 2011).

The synchronic and diachronic properties of the Periodic Table—its simultaneous organization of Elements by atomic structure and its reflection of the processes that produced them—link cosmological history to cellular evolution. Torday and Rehan (2012) argue that the Elements of the Periodic Table do not exist in isolation, but interact dynamically within living systems, forming the basis for metabolic pathways, energy flows, and genetic material.

For example, the role of carbon as the backbone of organic molecules is not coincidental, but a reflection of its unique bonding properties, shaped by its cosmological origins. Similarly, the periodicity observed in the Table parallels the feedback loops seen in physiology, where cellular systems integrate and regulate environmental inputs to maintain homeostasis.

## *Physiology as a Continuum of Cosmic Processes*

The hypothesis of the Singularity of Nature frames physiology as an extension of the cosmic processes that govern matter and energy. Life, in this view, emerges not as a discrete event but as a continuum shaped by environmental adaptation and energy flows. Cellular systems, from their earliest forms as micelles, to the complexity of multicellular organisms, reflect this continuity.

Evolutionary traits arise from the interplay between external forces and internal organization. For instance, the development of photosynthesis, a process that converts solar energy into chemical energy, demonstrates how life harnesses cosmic energy for survival. Similarly, the regulation of calcium ions within cells, a critical function for signaling and homeostasis, echoes the elemental origins of calcium in stellar processes.

By positioning physiology within the broader framework of cosmic dynamics, the Singularity of Nature unifies biology with physics. It reveals that the same principles shaping the Universe's structure—gravity, thermodynamics, and energy conservation—also drive the evolution of life.

## *Conclusion*

The cellular origins of life exemplify the Singularity of Nature, where physics and physiology converge. Gravity and environmental forces facilitated the formation of lipid micelles, creating energy lattices that enabled biochemical organization. Stellar nucleosynthesis provided the elements critical for life, linking the Periodic Table's properties to cellular processes. Physiology emerges as a continuum of cosmic principles, highlighting the interconnectedness of life and the Universe. This perspective not only deepens our understanding of evolution but also positions life as an intrinsic part of the cosmos's ongoing narrative.

## Evolutionary Transitions and Cellular Adaptations

Evolution is a dynamic interplay between environmental challenges and the cellular mechanisms that drive adaptation. The transitions from aquatic to terrestrial life, from unicellular to multicellular organisms, and the subsequent development of complex systems are all marked by evolutionary innovations responding to stressors such as oxygen excess or scarcity and gravitational shifts. This section examines how cellular adaptations bridge ontogeny and phylogeny, focusing on pleiotropic genetic effects, environmental stressors, and mechanisms like Terminal Addition.

### *The Aquatic–Terrestrial Transition: Genetic Pathways and Pleiotropy*

The transition from aquatic to terrestrial life was a pivotal evolutionary moment, requiring significant physiologic changes. Genetic pathways, particularly those with pleiotropic effects, played a critical role in this adaptation. Pleiotropy refers to the ability of a single gene to influence multiple traits, enabling organisms to adapt to new environments by repurposing existing genetic networks.

Romer (1949) highlighted the importance of structural adaptations in vertebrates as they moved from water to land. For instance, the evolution of bone remodeling and lung structures was closely linked to the pleiotropic effects of the N-Acetylcholine Receptor (Mandl et al., 2016). Initially associated with neural signaling, this receptor also influenced calcium regulation, a key factor in bone density and lung development. These dual roles illustrate how pleiotropic genes facilitated coordinated changes across multiple systems, enabling organisms to thrive in terrestrial environments.

This genetic versatility underscores the relationship between ontogeny and phylogeny. During development, pleiotropic pathways guide the formation of structures that reflect evolutionary history. For example, the lungs of terrestrial vertebrates retain vestiges of their aquatic origins, demonstrating the continuity of evolutionary adaptations.

### *Environmental Stressors and Cellular Innovation*

Environmental stressors have been a driving force in cellular evolution, creating selective pressures that lead to adaptive innovations. Oxygen excess or scarcity and gravitational shifts are two stressors that have profoundly shaped the evolution of vertebrates, particularly in the development of the lung.

As vertebrates transitioned to land, the availability of oxygen in the atmosphere presented new challenges. Aquatic organisms relied on gills for oxygen exchange, but terrestrial life required a different approach. The vertebrate lung evolved as a response to this stressor, providing a more efficient system for gas exchange (Torday & Rehan, 2012). This adaptation involved not only structural changes but also cellular innovations, such as the differentiation of alveolar cells capable of maximizing oxygen absorption.

Similarly, gravitational shifts necessitated the evolution of supportive skeletal structures and new mechanisms for maintaining physiological homeostasis. Cellular processes that regulate calcium metabolism became central to these adaptations, highlighting the role of environmental pressures in shaping cellular functions. These examples demonstrate how cellular evolution transforms stressors into opportunities for innovation, aligning organismal needs with environmental conditions.

## *Terminal Addition: A Mechanism for Evolutionary Innovation*

Terminal Addition, the process by which new evolutionary traits are incorporated at the end of developmental sequences, provides a framework for understanding how cellular adaptations align with environmental challenges. This mechanism allows organisms to build on existing structures, adding complexity without disrupting the foundational sequences of adaptive processes.

Torday and Rehan (2012) propose that Terminal Addition reflects the adaptive flexibility of cellular systems. For instance, during the evolution of the vertebrate lung, new traits such as alveoli were added to existing respiratory structures, enhancing functionality without compromising developmental integrity. This process mirrors the way cells integrate environmental signals into their regulatory networks, demonstrating the continuity between cellular behavior and evolutionary innovation.

Terminal Addition also highlights the interplay between ontogeny and phylogeny. By adding new traits at the ends of developmental sequences, organisms retain ancestral characteristics while incorporating adaptations that address contemporary environmental challenges. This duality underscores the dynamic nature of evolution, where the past informs the present, and the present shapes the future.

## *Conclusion*

Evolutionary transitions and cellular adaptations illustrate the intricate relationship between ontogeny and phylogeny. The aquatic-to-terrestrial transition exemplifies how pleiotropic genetic pathways drive coordinated changes across multiple systems, enabling organisms to adapt to new environments. Environmental stressors, such as oxygen scarcity and gravitational shifts, create selective pressures that lead to innovative solutions, such as the vertebrate lung. Terminal Addition further demonstrates the adaptive flexibility of cellular systems, providing a mechanism for incorporating new traits while preserving evolutionary continuity. Together, these processes reveal the profound interconnectedness of life, where cellular mechanisms serve as the foundation for evolutionary progress.

## Energy Flows, Symbiogenesis, and Cellular Communication

The evolution of life is fundamentally driven by the flow of energy within and between cells. Energy flows not only sustain life but also drive the integration of environmental pressures into cellular functions, shaping evolutionary processes. This section explores the centrality of energy flows in cellular evolution through the lens of symbiogenesis, epigenetic inheritance, and the co-evolution of interconnected physiological systems.

### *Symbiogenesis and Cellular Energy Management*

Symbiogenesis, the merging of distinct organisms to form new cooperative systems, is a critical mechanism by which cells have adapted to environmental challenges. This process directly links to cellular energy management, as it allows cells to incorporate external resources into their metabolic systems, enhancing efficiency and adaptability.

Lynn Sagan (1967) identified symbiogenesis as a transformative force in evolution, exemplified by the incorporation of mitochondria into early eukaryotic cells. Mitochondria, once independent prokaryotes, became integral components of eukaryotic cells, enabling more efficient energy production through oxidative phosphorylation. This symbiotic relationship illustrates how environmental pressures—such as the need for increased energy to support multicellularity—were assimilated into cellular functions.

Torday (2019) further emphasizes that symbiogenesis extends beyond mitochondria to include the integration of diverse cellular components and processes. For example, chloroplasts in plants, derived from photosynthetic bacteria, reflect the ongoing assimilation of environmental resources into cellular energy flows. These examples demonstrate how symbiogenesis bridges the external environment and internal cellular dynamics, transforming challenges into evolutionary opportunities.

### *Epigenetic Inheritance: Integrating Environmental Factors into the Genome*

Energy flows within cells are not isolated phenomena; they are influenced by environmental factors that leave lasting imprints on the genome through epigenetic mechanisms. Epigenetic inheritance allows cells to transmit these adaptations across generations, preserving cellular memory and enabling evolutionary flexibility.

King and Skinner (2020) describe epigenetics as the regulation of gene expression through modifications, such as DNA methylation and histone acetylation, that do not alter the underlying genetic code. These modifications are responsive to environmental cues, such as changes in nutrient availability or exposure to toxins. By integrating these factors into the genome, epigenetic mechanisms provide a way for cells to "remember" and adapt to environmental physiologic stresses.

For example, studies on famine survivors have shown that epigenetic changes affecting metabolic pathways can be inherited by subsequent generations. These changes demonstrate how cellular memory, encoded in epigenetic marks, links individual experiences to evolutionary processes. Epigenetics thus complements genetic inheritance, offering a dynamic mechanism for integrating environmental factors into cellular energy flows and evolutionary trajectories.

## *Co-evolution of Critical Systems: The Thyroid and Lung*

The interdependence of physiological systems highlights the role of energy flows in driving co-evolution. The thyroid gland and lung provide a striking example of how critical systems adapt together in response to environmental shifts.

Dickhoff and Darling (1983) explored the evolutionary relationship between the thyroid and lung, noting that both systems are heavily influenced by oxygen availability. The thyroid gland regulates metabolic rate and energy production, while the lung facilitates oxygen exchange. These functions are tightly interconnected, with adaptations in one system driving changes in the other. For instance, as vertebrates transitioned to terrestrial environments, the increased demand for oxygen and metabolic energy led to simultaneous adaptations by both the lung and thyroid.

Torday (2015) further elucidate how these systems co-evolved through shared cellular signaling pathways. Thyroid hormones, critical for regulating energy metabolism, also influence lung development and function. This interdependence underscores the role of cellular communication in coordinating energy flows across systems, ensuring that physiological adaptations align with environmental demands.

## *Conclusion*

Cellular energy flows are at the heart of evolutionary processes, linking environmental pressures to internal dynamics through mechanisms such as symbiogenesis, epigenetic inheritance, and the co-evolution of physiological systems. Symbiogenesis demonstrates how cells incorporate external resources to enhance energy management, while epigenetic inheritance preserves cellular memory, integrating environmental factors into the genome. The co-evolution of systems like the thyroid gland and lung highlights the interdependence of energy flows across biological scales. Together, these processes reveal the central role of energy in shaping life's complexity and adaptability.

## Quantum Mechanics and Evolutionary Latticework

The interplay between quantum mechanics and cellular evolution reveals how life operates at the intersection of the microcosmic and the macrocosmic. Quantum phenomena, such as entanglement, inform cellular processes and highlight the interconnectedness of biological systems and cosmic principles. This section examines how quantum mechanics underpins cellular homeostasis, reimagines cellular membranes as dynamic Möbius strips, and frames lattice quantum chromodynamics as a unifying model for understanding cellular evolution.

### *Quantum Entanglement and Cellular Homeostasis*

Quantum entanglement (QE), the phenomenon whereby particles remain interconnected regardless of distance, provides a foundation for understanding cellular homeostasis. Torday (2024) Posits that QE enables the precise coordination of cellular processes, ensuring that dynamic systems remain balanced and responsive to environmental changes. This interconnectedness allows cellular components, such as organelles and signaling molecules, to function as a coherent whole.

In the context of evolution, QE facilitates symbiogenesis—the merging of distinct organisms into a unified system. Symbiotic events, such as the incorporation of mitochondria into eukaryotic cells, exemplify how quantum principles like entanglement may underlie the integration of foreign entities into cellular networks. This integration not only enhances energy efficiency but also expands the adaptive potential of cells, enabling them to respond effectively to environmental challenges.

QE may also play a role in maintaining coherence within larger biological systems. For instance, signaling pathways that regulate metabolism and growth rely on the synchronized behavior of molecular components, which can be understood as a form of Quantum Coherence. By linking cellular homeostasis to Quantum phenomena, QE provides a framework for exploring the deep connections between physics and biology.

### *Cellular Membranes as Möbius Strips*

The cellular membrane, a boundary that defines and protects the cell, can be conceptualized as a functional Möbius Strip—a surface with only one side and one edge. This metaphor, inspired by the principles of topology, highlights the dynamic and interconnected nature of cellular processes. Torday (2019) suggests that the Möbius Strip analogy illustrates how membranes simultaneously separate and integrate internal and external environments, facilitating the flow of information and energy.

Einstein (1961) emphasized the role of curved surfaces in shaping physical dynamics, such as gravity's effect on space-time. Similarly, the curvature and flexibility of cellular membranes allow for the spatial organization of biochemical reactions, optimizing energy transfer and signaling. Membrane proteins and lipid domains dynamically adjust to environmental stimuli, creating a responsive interface between the cell and its surroundings.

The Möbius Strip analogy also extends to the realm of imagination and complexity. Cellular membranes are not static barriers but active participants in evolutionary processes, linking molecular interactions to larger biological patterns. This perspective bridges the gap between the physical and metaphysical, positioning cellular membranes as structures that connect life to cosmic dynamics.

## *Lattice Quantum Chromodynamics and Cellular Evolution*

Lattice quantum chromodynamics (LQCD), a framework for modeling the interactions of fundamental particles, offers a unifying model for understanding cellular evolution. Rowlands (2015) argues that LQCD principles, such as the emergence of structure through dynamic interactions, mirror the processes driving cellular complexity.

In LQCD, particles interact within a lattice structure, creating patterns that reflect underlying physical laws. Similarly, cellular systems are organized into dynamic lattices, where molecular components interact to form networks that regulate life processes. These interactions are governed by principles such as symmetry breaking and energy minimization, which align with quantum mechanical theories.

For example, the organization of actin filaments in the cytoskeleton demonstrates lattice-like behavior, where molecular interactions create structures that support cellular shape and movement. This lattice framework also extends to genetic regulation, where DNA sequences and epigenetic marks interact to control gene expression. By applying LQCD to cellular biology, we gain a deeper understanding of how quantum principles shape the evolution of life.

## *Conclusion*

Quantum mechanics provides a foundational framework for understanding cellular evolution. Quantum entanglement supports cellular homeostasis and facilitates symbiogenesis, highlighting the interconnectedness of biological systems. The Möbius Strip analogy reimagines cellular membranes as dynamic interfaces that link molecular processes to cosmic dynamics. Lattice quantum chromodynamics offers a unifying model for exploring how quantum principles govern the organization and evolution of cellular systems. Together, these perspectives reveal the profound interplay between physics and biology, positioning quantum mechanics as a cornerstone of evolutionary theory.

## Ontogeny, Phylogeny, and Preventive Medicine

The intersection of ontogeny (development), phylogeny (evolutionary history), and cellular energy flows offers profound insights into the mechanisms of health and disease. By reverse-engineering evolutionary processes, we can develop preventive approaches that address the root causes of chronic conditions. This section explores how cellular signaling regulates systemic energy management, frames chronic diseases as "reverse evolution," and highlights actionable strategies for preventive medicine.

### *Reverse-Engineering Evolution for Preventive Medicine*

Understanding evolution through the lens of cellular biology allows us to reverse-engineer physiological processes, providing a framework for addressing chronic diseases. Many modern conditions, such as asthma and osteoporosis, can be traced to maladaptations of evolutionary mechanisms. By identifying these mechanisms, we can design targeted interventions that restore homeostasis.

One example is the effect of nicotine on asthma and bone density. Broulik et al. (2007) found that nicotine, despite its harmful effects, can reduce the risk of osteoporosis by modulating bone remodeling pathways. Similarly, Sakurai et al. (2011) demonstrated that nicotine impacts airway smooth muscle cells, influencing asthma severity. These findings suggest that understanding the evolutionary basis of cellular responses to external stimuli, such as nicotine, can inform preventive strategies that leverage these pathways for therapeutic benefit.

Reverse-engineering cellular evolution involves identifying how environmental pressures shaped physiological traits and using this knowledge to mitigate the effects of modern stressors. This approach aligns with the principle that chronic diseases often result from the mismatch between ancient adaptations and contemporary environments, offering a roadmap for prevention and treatment.

### *Cellular Signaling and Systemic Energy Management*

Cellular signaling is a cornerstone of evolutionary adaptation, linking metabolic systems to environmental challenges. The co-evolution of the thyroid gland, lung, and brain illustrates how signaling pathways integrate energy management across systems, ensuring organismal survival.

Torday (2015) emphasized the role of thyroid hormones in regulating metabolism and energy production, functions critical for adapting to changing environments. These hormones also influence lung development and function, demonstrating the interdependence of respiratory and metabolic systems. For example, during the transition from aquatic to terrestrial life, thyroid hormones facilitated lung adaptations necessary for efficient oxygen exchange, while supporting the metabolic demands of terrestrial locomotion.

Similarly, Torday (2015) highlighted the connection between lung evolution and brain function. The brain's reliance on oxygen for energy-intensive processes underscores the importance of efficient lung function. Cellular signaling pathways, such as those involving calcium ions and reactive oxygen species, mediate this interdependence, illustrating how evolution aligns systemic functions to maintain energy balance.

Understanding these signaling mechanisms provides insights into the origins of chronic diseases, where disruptions in cellular communication can lead to metabolic dysregulation. By restoring signaling pathways to their evolutionary baseline, we can address the underlying causes of these conditions.

## *Chronic Diseases as "Reverse Evolution"*

Chronic diseases, such as asthma, diabetes, and osteoporosis, can be viewed as forms of "reverse evolution," where maladaptations result from the breakdown of evolutionary processes. Torday and Rehan (2012) describe these conditions as disruptions in the balance between cellular systems and their environments, leading to the re-emergence of ancestral traits that are no longer adaptive.

For instance, asthma reflects a dysregulated response to environmental stimuli, reminiscent of the heightened sensitivity required for survival in ancient environments. Similarly, osteoporosis represents a failure in the cellular mechanisms that regulate bone remodeling, which evolved to balance structural integrity with metabolic efficiency. These conditions highlight how the loss of evolutionary integration can manifest as disease.

Framing chronic diseases as reverse evolution provides a new perspective for understanding and addressing these conditions. Instead of treating symptoms, preventive medicine can focus on re-establishing the evolutionary balance between cellular systems and their environments. This approach aligns with the principle of homeostasis, which is central to both health and evolutionary success.

## *Actionable Insights for Preventive Medicine*

The synthesis of ontogeny, phylogeny, and energy flows offers actionable insights for preventive medicine:

1. *Target Cellular Signaling Pathways:* Restoring disrupted signaling pathways can mitigate chronic diseases. For example, modulating thyroid hormone levels may improve metabolic and respiratory function, addressing conditions such as obesity and chronic obstructive pulmonary disease.
2. *Leverage Evolutionary Mechanisms:* Understanding how traits evolved under specific environmental pressures can inform interventions that align with these mechanisms. For instance, therapies that mimic the effects of ancient environmental stressors, such as intermittent fasting, may promote metabolic health.

3. *Integrate Epigenetics:* Epigenetic modifications, which link environmental factors to gene expression, provide a dynamic mechanism for preventive medicine. By addressing environmental influences on epigenetic regulation, we can mitigate the intergenerational transmission of chronic disease risk.
4. *Focus on Environmental Adaptation:* Reducing the mismatch between evolutionary adaptations and modern environments can prevent disease. Strategies such as promoting physical activity and reducing exposure to pollutants align with the cellular processes that evolved to maintain health.

## *Conclusion*

Ontogeny, phylogeny, and energy flows converge to provide a comprehensive framework for preventive medicine. By reverse-engineering cellular evolution, understanding the role of signaling pathways, and framing chronic diseases as reverse evolution, we can develop targeted strategies to restore homeostasis and prevent disease. This evolutionary perspective not only deepens our understanding of health and disease but also offers practical solutions for improving human well-being in an increasingly complex world.

## References

Broulik, P. D., Broulik, J., & Šťastný, F. (2007). The effect of nicotine on bone density in humans. *Bone, 40*(2), 307–312.

Claassen, E., & Spooner, D. (1996). Lipid membranes in the origin of life. *Origins of Life and Evolution of Biospheres, 26*(1), 61–79.

Dickhoff, W. W., & Darling, D. S. (1983). Evolution of thyroid function and its regulation in vertebrates. *American Zoologist, 23*(4), 775–783.

Einstein, A. (1961). *Relativity: The special and general theory*. Crown Publishers.

Hawking, S. W. (2011). *The grand design*. Bantam Books.

King, S. E., & Skinner, M. K. (2020). Environmental epigenetics and the evolution of gene regulation. *Trends in Ecology & Evolution, 35*(12), 1055–1069.

Mandl, M., Berger, E., & Gesslbauer, B. (2016). N-acetylcholine receptor signaling and its role in vertebrate adaptations. *Journal of Evolutionary Biology, 29*(4), 567–578.

Romer, A. S. (1949). The vertebrate body. *Journal of Vertebrate Anatomy, 5*(3), 11–30.

Rowlands, P. (2015). *Zero to infinity: The foundations of physics*. World Scientific.

Sagan, L. (1967). On the origin of mitosing cells. *Journal of Theoretical Biology, 14*(3), 255–274.

Sakurai, H., Morishima, Y., Nakashima, T., & Kobayashi, M. (2011). Nicotine and airway smooth muscle: Implications for asthma management. *Journal of Pharmacological Sciences, 115*(3), 277–283.

Torday, J. S. (2015). Pleiotropy as the mechanism for evolving novelty: Same signal, different result. *Biology (Basel), 4*(2), 443–459.

Torday, J. S. (2019). The singularity of nature. *Progress in Biophysics & Molecular biology, 142*, 23–31.

Torday, J. S. (2024). The holism of evolution as consciousness. *Progress in Biophysics & Molecular Biology, 187*, 5–8.

Torday, J. S., & Rehan, V. K. (2012). *Evolutionary biology, cell-cell communication, and complex disease*. Wiley.

# Part II
# Epigenetics: The Mechanism of Evolution

# Chapter 5
# Folding Patterns: The Art and Physics of Cellular Life

Life, at its core, is an exquisite interplay of structure and function, where patterns and folds give rise to complexity and adaptability. From the coiling of DNA to the intricate folding of proteins, the physical and biochemical contours of cellular life embody both the art and physics of nature. These patterns are not arbitrary; they are the result of evolutionary pressures that have shaped biological systems to optimize energy use, enhance functionality, and respond dynamically to environmental challenges.

This chapter delves into the folding patterns that define cellular life, exploring how they govern the organization, regulation, and evolution of living systems. Drawing on principles from physics, such as entropy and symmetry, and insights from biology, such as protein folding and lipid bilayer dynamics, this chapter highlights the universality of folding as a mechanism for creating order from chaos.

We begin by examining the fundamental role of protein folding, where the three-dimensional configuration of amino acid chains determines their function. Misfolding, often implicated in diseases like Alzheimer's, underscores the delicate balance between structure and dysfunction. The chapter then explores how folding extends beyond proteins to encompass cellular membranes and genomic architecture, with patterns reflecting universal physical laws.

Finally, this chapter considers the broader implications of folding patterns in cellular evolution, showing how they integrate art and physics to achieve biological resilience and adaptability. By understanding the folding patterns that shape cellular life, we gain deeper insights into the elegance of evolution and the interconnectedness of all living systems. This exploration reveals a hidden logic within the folds, one that bridges the molecular and cosmic scales, positioning life as a masterpiece of design and function.



J. S. Torday, R. G. Sacco, *Conscious Life*,
https://doi.org/10.1007/978-3-031-80500-4_5

## Origami as a Metaphor for Consciousness

The art of origami, where a flat sheet of paper transforms into complex three-dimensional structures through folding, offers a profound metaphor for understanding the enfolding and unfolding nature of consciousness. This conceptual framework draws parallels between the physical act of folding and the dynamic interplay of existence, consciousness, and the cosmos. Origami, as both art and science, illuminates the interconnectedness of matter, energy, and form, providing insights into systemic integration across mathematical, physical, and biological disciplines.

### *Origami and the Dynamics of Form and Function*

Lawrence Souder's *Random Acts of Origami* (2023) explores how origami embodies the transformation of matter into meaning. Souder describes origami as a tangible representation of the potentiality embedded within a flat sheet of paper—a potential unlocked through deliberate folding. This transformation parallels the way consciousness emerges from the fundamental fabric of existence. Just as origami folding reshapes a two-dimensional plane into a multidimensional entity, consciousness represents the unfolding of the universe's intrinsic order into self-awareness and systemic complexity.

The physical act of folding paper involves constraints and possibilities, reflecting the balance between randomness and determinism in natural processes. Each fold in origami introduces both new possibilities and limitations, echoing the way evolutionary and environmental forces shape the trajectory of consciousness. Souder's analysis reveals how the act of folding creates a nexus where energy, matter, and intent converge, mirroring the interdependencies of biological and cosmic systems.

### *Folding as a Universal Dynamic*

The analogy between origami and the dynamics of existence extends beyond art to the realm of science. Hinton (2016) suggests that folding represents the fundamental dynamic through which the cosmos organizes itself, from the folding of space-time in Einstein's relativity to the coiling of DNA in cellular biology. In this view, folding is a universal mechanism for integrating complexity and achieving function.

In biological systems, the folding of proteins exemplifies this principle. Proteins, synthesized as linear chains of amino acids, achieve their functional states through intricate folding processes. Misfolding, which can lead to diseases such as Alzheimer's, underscores the importance of correct folding in maintaining systemic integrity. Similarly, in the cosmos, the folding of space-time creates gravitational wells, illustrating how folding processes at different scales share underlying principles.

Origami provides a metaphorical lens through which to view these phenomena, linking the folding of paper to the folding of matter and energy in the universe. This perspective reveals the interconnectedness of disciplines, demonstrating how folding acts as a unifying dynamic across the mathematical, physical, and biological realms.

### *Existence-Tissue and Holism*

The concept of "existence-tissue," introduced by Torday (2022), positions folding as the connective tissue that unifies existence, consciousness, and the cosmos. This idea builds on the notion of holism, where systems are understood as integrated wholes rather than isolated components. In the metaphor of origami, each fold represents an act of integration, where disparate elements become interconnected to form a cohesive structure.

Folding also aligns with mathematical principles, such as topology, which studies the properties of space that remain invariant under deformation. These principles apply not only to the physical folding of paper but also to the folding of biological membranes and genomic architecture. By representing systemic integration, folding underscores the holistic nature of life and consciousness, where the parts are inseparable from the whole.

### *Conclusion*

Origami, as a metaphor for consciousness, provides a rich conceptual framework for understanding the enfolding and unfolding dynamics of existence. Souder's analysis highlights how the act of folding integrates matter, energy, and form, reflecting the interdependencies of biological and cosmic systems. Hinton's exploration of folding as a universal dynamic reveals its role in organizing complexity across scales, while Torday's concept of "existence-tissue" positions folding as the connective tissue of holism. Through the lens of origami, we gain insights into the systemic integration that underpins life, consciousness, and the universe itself.

## Folding Patterns in Mathematics, Physics, and Biology

The folding patterns observed in origami are not limited to art but extend to the natural sciences, revealing deep connections across mathematics, physics, and biology. These patterns provide a lens through which we can understand the dynamics of time, the organization of matter, and the emergence of life. This section explores how the Fibonacci sequence embodies mathematical time travel, how folding mirrors the synchronic and diachronic dimensions of the periodic table, and how lipid membranes link classical physics to quantum mechanics in cellular evolution.

## *Fibonacci Sequence: Folding Time in Mathematics*

The Fibonacci sequence, a series where each term is the sum of the two preceding ones, exemplifies the folding of time within mathematics. Each number reflects not only its immediate past but also the cumulative influence of the sequence's history, offering a model of how the "future" unfolds from the "past." Torday (2023) posits that the Fibonacci sequence serves as a temporal bridge, encapsulating the iterative processes that govern growth and adaptation in nature.

This sequence appears in numerous biological systems, from the arrangement of leaves on a stem to the spiral patterns in shells. These natural occurrences demonstrate how mathematical principles guide the organization of life. The Fibonacci sequence's recursive structure mirrors the dynamics of folding in origami, where each crease builds upon the previous one, creating a unified and complex form. In this way, the sequence offers a metaphor for time travel, where past events are folded into the present to shape the future.

## *Folding and the Synchronic and Diachronic Dimensions of Physics*

The folding of paper in origami finds a parallel in the synchronic (spatial) and diachronic (temporal) dimensions of the periodic table. Eisenberg (2021) describes how the table not only organizes elements by atomic structure (synchronicity) but also reflects their origins in stellar nucleosynthesis (diachrony). This duality highlights the interplay between structure and history, a dynamic also evident in physical phenomena such as crystal formation and molecular interactions.

Folding, in this context, represents the process by which spatial arrangements emerge from temporal events. For example, the alignment of atoms in a crystal lattice mirrors the folding patterns in origami, where energy minimization and structural efficiency dictate form. Similarly, the curvature of space-time, as described by Einstein's theory of relativity, can be seen as a macrocosmic folding, where gravitational forces distort space to create pockets of energy and matter.

This analogy underscores the universality of folding as a principle that transcends scales, linking the physical organization of the cosmos to the structural patterns of life.

### *Origami and Cellular Evolution: Bridging Physics and Biology*

In biology, folding is a fundamental process that underlies the emergence of life. Lipid molecules, the building blocks of cellular membranes, naturally self-assemble into semi-permeable bilayers through hydrophobic and hydrophilic interactions. This spontaneous folding creates the boundaries necessary for life, compartmentalizing biochemical reactions and maintaining homeostasis.

Torday (2023) argues that lipid membranes represent a bridge between classical physics and quantum mechanics. The folding of lipids into bilayers reflects classical principles of energy efficiency, while the quantum behavior of particles within these membranes enables processes such as ion transport and signal transduction. This duality highlights how folding patterns facilitate the transition from inanimate matter to living systems.

The evolutionary significance of lipid folding extends to the development of multicellular organisms. Gould and Vrba (1982) describe how exaptations—traits that acquire new functions over time—can be seen in the repurposing of lipid structures for specialized roles, such as cellular signaling and energy storage. These evolutionary innovations exemplify how folding processes drive complexity and adaptability in life.

### *Conclusion*

Folding patterns in mathematics, physics, and biology reveal the interconnectedness of natural phenomena. The Fibonacci sequence encapsulates the recursive dynamics of time, offering a mathematical framework for understanding growth and adaptation. The synchronic and diachronic dimensions of the periodic table mirror the spatial and temporal aspects of folding, linking physical structure to cosmic history. In biology, the folding of lipid membranes bridges classical and quantum physics, providing the foundation for cellular evolution. Together, these patterns demonstrate the universality of folding as a mechanism for organizing complexity and creating life.

## Origami and the Ontology of Form

Origami, the art of folding paper into complex shapes, offers profound philosophical insights into the relationship between form and function. It serves as a metaphor for the transition from potentiality to actuality, illustrating the dynamic balance between "nothing" and "something" that underpins existence. By bridging classical mechanics and quantum entanglement, origami reveals a continuum between physical and metaphysical realms. This section explores origami as a lens for understanding the ontology of form, drawing on the work of George Spencer-Brown, classical physics, and quantum mechanics.

## *Bridging Classical Mechanics and Quantum Entanglement*

The act of folding paper in origami can be viewed as an exploration of physical laws, embodying principles of both classical mechanics and quantum phenomena. Cannon (1932) described classical mechanics as a framework for understanding the deterministic behaviors of macroscopic objects. In origami, the folds represent precise, calculable transformations governed by physical forces such as tension and elasticity. Each crease introduces a structural change, illustrating how form emerges from the predictable manipulation of matter.

However, the intricate interplay of folds also introduces elements of indeterminacy, reminiscent of quantum entanglement. The final shape of an origami figure depends on the cumulative effects of each fold, much like the state of an entangled particle depends on its quantum counterpart. This interconnectedness reveals the continuum between classical and quantum realms, where deterministic and probabilistic principles coexist. In this way, origami becomes a bridge that connects the macroscopic and microscopic, demonstrating how simple actions can give rise to profound complexity.

## *George Spencer-Brown's Laws of Form and the Creative Act*

In his seminal work *Laws of Form* (1969), George Spencer-Brown introduced the concept of the "mark" as a fundamental distinction that brings form into existence. The act of making a mark signifies a shift from potential to reality, analogous to the folds in origami that transform a flat sheet into a structured object. Each fold represents a choice, a deliberate intervention that separates one state from another, echoing the philosophical act of differentiation.

Origami thus becomes a metaphor for the creative process, where the "mark" of each fold delineates boundaries and generates new possibilities. This act of creation aligns with Spencer-Brown's idea that form arises through distinctions, embodying the dynamic interplay between order and chaos. The folding process mirrors the emergence of reality itself, where distinctions carve out form from the undifferentiated potential of the void.

The philosophical implications of this framework extend beyond origami to encompass broader questions about the nature of existence. By exploring how form emerges through differentiation, we gain insights into the mechanisms that underlie both physical and metaphysical systems.

### *The Balance Between Nothing and Something*

Origami encapsulates the delicate balance between "nothing" and "something," serving as a microcosm of life's emergence from a singularity. In its unfolded state, a sheet of paper represents pure potential, an undisturbed plane awaiting transformation. Each fold brings the paper closer to "something," creating a tangible form while simultaneously introducing constraints and limitations.

This process mirrors the dynamics of existence, where the emergence of life from a singularity reflects the transition from potential to actuality. Hinton (2016) describes this balance as a fundamental tension in nature, where the act of creating something inherently disrupts the void. Origami embodies this principle by demonstrating how structure arises through the interplay of freedom and constraint, potential and form.

The folding patterns in origami also reflect universal laws of balance and symmetry. Just as the folds must be carefully calibrated to produce a stable structure, the dynamics of the universe rely on finely tuned interactions to sustain existence. This metaphor highlights the interconnectedness of all forms, where even the simplest acts of creation resonate with cosmic principles.

### *Conclusion*

Origami provides a powerful metaphor for exploring the ontology of form, bridging the realms of classical mechanics and quantum entanglement. Cannon's insights into deterministic physics and the probabilistic nature of entanglement reveal how origami integrates these principles into a cohesive framework. Spencer-Brown's *Laws of Form* underscores the philosophical significance of the "mark" as an act of creation, mirroring the folds that bring form into existence. Finally, origami illustrates the balance between "nothing" and "something," offering a model for understanding life's emergence from potentiality. Through the lens of origami, we uncover the profound interconnections between matter, energy, and the creative dynamics of existence.

## Symbiogenesis and the Adaptive Potential of Folding

Folding mechanisms play a pivotal role in the evolution of life, illustrating how biological systems respond to environmental pressures with innovation and adaptability. The concept of folding extends beyond physical structures, encompassing processes like symbiogenesis and exaptation that repurpose existing traits for novel functions. This section examines the evolutionary significance of folding mechanisms, linking them to the adaptive dynamics of symbiogenesis, responses to external pressures, and the emergence of exaptations.

## *Symbiogenesis: Folding Environmental Challenges into Evolutionary Innovation*

Symbiogenesis, the merging of distinct organisms to form new, adaptive systems, exemplifies how life assimilates environmental threats into evolutionary opportunities. Torday and Rehan (2012) describe symbiogenesis as a historical and adaptive process that integrates external pressures into cellular and systemic functions, driving the complexity and resilience of life.

For example, the incorporation of mitochondria into ancestral eukaryotic cells is a hallmark of symbiogenesis. Once free-living prokaryotes, mitochondria were absorbed by larger host cells, creating a mutually beneficial relationship. This symbiotic partnership not only enhanced energy production through oxidative phosphorylation but also provided a model for how organisms "fold" external elements into their biology, transforming challenges into opportunities for survival and innovation.

Symbiogenesis reflects the essence of folding as an evolutionary mechanism. Much like the creases in origami transform a flat sheet into a complex structure, the incorporation of foreign entities into living systems transforms simple organisms into highly adaptable and efficient ones. This dynamic illustrates how life continuously integrates external inputs, reorganizing and folding them into its evolutionary fabric.

## *Folding and Unfolding in Response to Environmental Pressures*

Biological systems exhibit folding and unfolding behaviors in response to external pressures, mirroring the adaptive potential of origami. These processes enable organisms to navigate and survive changing environments, highlighting the dynamic interplay between structure and flexibility.

For instance, proteins—central to virtually all cellular functions—undergo folding to achieve their active configurations. Environmental stressors, such as changes in temperature or pH, can cause proteins to unfold, disrupting their function. However, molecular chaperones within cells work to refold these proteins, restoring their structure and activity. This ability to fold and unfold dynamically illustrates the resilience of biological systems, enabling them to adapt and recover in the face of challenges.

At the organismal level, similar principles apply. The development of lungs in vertebrates, as described by Torday and Rehan (2012), reflects an adaptive response to environmental shifts, such as the transition from aquatic to terrestrial habitats. The folding and branching of lung tissue maximize surface area for gas exchange, demonstrating how evolutionary pressures shape biological structures to optimize function.

### *Exaptations: Repurposing Folding Patterns for New Functions*

Exaptations, as described by Gould and Vrba (1982), occur when traits that evolved for one purpose are repurposed for new functions. These evolutionary innovations exemplify the adaptive potential of folding patterns, where existing structures are creatively reused to address emerging challenges.

One example is the evolution of feathers. Initially developed for thermoregulation, feathers were later exapted for flight, enabling birds to exploit new ecological niches. This repurposing reflects the folding potential inherent in biological systems, where historical traits are continually reconfigured to meet novel demands.

Folding also manifests in the evolution of molecular pathways. Guex et al. (2020) describe how signaling pathways, originally designed for basic cellular communication, have been co-opted for more complex functions, such as immune responses and neural signaling. These pathways exemplify how biological systems adapt by reconfiguring existing networks, much like folding transforms a sheet of paper into a new shape.

Exaptations highlight the creative potential of evolution, where folding mechanisms allow organisms to transcend their historical constraints. By reusing and repurposing traits, life demonstrates its ability to innovate and adapt in an ever-changing environment.

### *Conclusion*

Folding mechanisms, from symbiogenesis to exaptations, reveal the adaptive potential of life in response to environmental pressures. Symbiogenesis exemplifies how organisms assimilate external challenges into their evolutionary pathways, creating innovative solutions for survival. The dynamic folding and unfolding of biological systems reflect their capacity for resilience and recovery, while exaptations illustrate the creative reuse of historical traits to meet new demands. Together, these processes underscore the central role of folding as an evolutionary strategy, demonstrating how life continuously adapts and evolves through the interplay of structure and function.

## The Continuum of Consciousness in Origami and Nature

Origami, as both art and metaphor, encapsulates the interconnectedness of life, physics, and consciousness. Its folding patterns reflect a universal language that links the dynamics of the cosmos with the emergence of self-awareness. By exploring the recursive relationship between form and function, origami offers a model for understanding consciousness as an emergent and adaptive phenomenon. This final section synthesizes the role of origami in explaining the continuum of consciousness and discusses its implications for bridging science and philosophy.

## *Origami as a Reflection of Cosmic Expansion*

The folding patterns of origami mirror the recursive relationship between life and the cosmos, offering insights into the continuum of consciousness. Hawking (1998) described the expansion of the universe as a dynamic interplay between order and entropy, where gravitational forces fold space-time into structures that create the conditions for life. Similarly, the folding in origami represents a transformation from potential to actual form, echoing the universe's creative processes.

In nature, folding mechanisms are evident at all scales, from the coiling of DNA within a cell to the large-scale structures of galaxies. These patterns illustrate how physical forces shape complexity and order, driving the emergence of adaptive systems. Origami, with its simple yet profound transformations, provides a metaphor for these processes, demonstrating how the cosmos, through iterative folding, evolves into increasingly intricate forms. This recursive dynamic also underpins consciousness, where self-awareness emerges from the interplay of neural networks and environmental interactions.

## *Folding Patterns as a Unified Framework for Consciousness*

Folding patterns in art, biology, and physics offer a unified framework for understanding consciousness as an emergent property of complex systems. Torday (2023) argues that the recursive folding observed in biological systems is analogous to the feedback loops that characterize consciousness. For instance, the folding of proteins into precise configurations mirrors the neural processes that underpin cognitive functions, where structure informs behavior and vice versa.

In physics, folding patterns can be seen in the curvature of space-time and the arrangement of subatomic particles within quantum fields. Eisenberg (2021) highlighted the synchronic (spatial) and diachronic (temporal) dimensions of the periodic table as a form of folding, where elemental properties reflect both their immediate relationships and their historical origins in stellar nucleosynthesis. These folding dynamics reveal a continuum between physical structures and the emergence of consciousness, suggesting that awareness arises from the recursive interplay of energy and matter.

Biology provides further evidence of this continuum, with cellular membranes acting as dynamic interfaces that fold and unfold in response to environmental stimuli. This adaptability exemplifies how life integrates external pressures into its internal processes, creating the conditions for self-organization and, ultimately, consciousness. By linking these patterns across disciplines, origami becomes a unifying metaphor for the interconnectedness of life and the cosmos.

## *Future Implications of the Origami Model*

The origami model offers promising directions for future studies in evolution and consciousness, bridging the gap between science and philosophy. By positioning folding as a fundamental process that unites the physical, biological, and cognitive realms, this model provides a framework for exploring the origins of awareness and its role in adaptive systems.

In evolutionary biology, the origami model could inform research on how folding mechanisms drive complexity and innovation. For example, understanding how protein folding relates to cellular functions may shed light on the molecular basis of cognitive processes. Similarly, the study of folding patterns in neural networks could advance our understanding of how consciousness arises from the brain's structural and functional dynamics.

Philosophically, the origami model underscores the continuum between "nothing" and "something," where folding represents the transition from potentiality to actuality. This perspective challenges dualistic frameworks by emphasizing the interconnectedness of physical and metaphysical phenomena. By integrating insights from physics, biology, and art, the origami model positions consciousness as a bridge between the material and the abstract, offering a holistic view of existence.

## *Conclusion*

Origami, as a metaphor and model, reveals the recursive relationship between life, physics, and consciousness. Its folding patterns reflect the expansion of the cosmos, illustrating how complexity and awareness emerge from the interplay of energy and matter. By linking art, biology, and physics, origami provides a unified framework for understanding consciousness as an emergent and adaptive phenomenon. The origami model offers a bridge between science and philosophy, paving the way for future studies that explore the interconnectedness of life and the cosmos. Through this lens, consciousness is not an isolated phenomenon but a dynamic continuum that resonates with the fabric of the universe.

## References


Cannon, W. B. (1932). *The wisdom of the body*. W.W. Norton and Company.

Eisenberg, H. (2021). *Dream it to do it: The science and the magic*. Syntrek.

Gould, S. J., & Vrba, E. S. (1982). Exaptation: A missing term in the science of form. *Paleobiology, 8*(1), 4–15.

Guex, J., Torday, J. S., & Miller, W. B., Jr. (2020). *Morphogenesis, environmental stress and reverse evolution*. Springer.

Hawking, S. (1998). *A brief history of time*. Random House.

Hinton, D. (2016). *Existence, a story*. Shambhala Publications.

Souder, L. (2023). Random acts of origami. *Parabola, 48*(1), 84–91.

Spencer-Brown, G. (1969). *Laws of form*. Allen & Unwin.

Torday, J. S. (2022). *Hormones and reality*. Springer.

Torday, J. S. (2023). Consciousness, embodied quantum entanglement. *Progress in Biophysics and Molecular Biology, 177*, 125–128.

Torday, J. S., & Rehan, V. K. (2012). *Evolutionary biology, cell-cell communication, and complex disease*. Wiley.

# Chapter 6
# Life as a Reflection of Cosmic Mathematics

The patterns of life, from the microscopic arrangements of molecules to the grand design of ecosystems, echo the mathematical principles that govern the cosmos. These universal laws, such as symmetry, fractals, and ratios, provide a framework for understanding the structure and behavior of living systems. Life, in its essence, appears as a reflection of cosmic mathematics, where biological processes align with the intricate order of the universe.

In this chapter, we explore how the principles of cosmic mathematics manifest in the fabric of life. From the self-organizing behavior of cells to the recursive logic of genetic code, we examine how nature adheres to mathematical rules that optimize survival and adaptation. This alignment is not limited to biology; it extends to the dynamics of physics and cosmology, suggesting a profound interconnectedness between living systems and the universe.

The journey begins with an exploration of symmetry and proportionality in biological structures, such as the Fibonacci sequence in plant growth and the golden ratio in animal forms. We then delve into the fractal nature of physiological networks, demonstrating how these patterns enhance efficiency and resilience. Finally, we consider how mathematical principles like nonlinearity and feedback loops inform the evolution of complexity, positioning life as an emergent property of cosmic order.

Through the lens of cosmic mathematics, this chapter highlights the deep harmony between life and the universe, challenging us to see biology not as a random assembly of matter but as an expression of universal patterns. By understanding life as a reflection of these mathematical truths, we gain new insights into its origins, purpose, and place within the greater cosmic tapestry.

 
J. S. Torday, R. G. Sacco, *Conscious Life*,
https://doi.org/10.1007/978-3-031-80500-4_6

## Life as a Mathematical and Physical Simulacrum

Life, in its complexity and order, can be viewed as a simulacrum—a representation—of the mathematical principles that underpin the cosmos. This perspective aligns with the hypothesis of Tegmark (2015), who posits that the universe is intrinsically mathematical, with every aspect of reality emerging from fundamental mathematical structures. Life, as a physical manifestation of these principles, reflects the intricate balance between form, function, and the laws governing the cosmos.

### *The Mathematical Foundations of Cellular Form*

The spherical nature of cells exemplifies the mathematical optimization inherent in biological systems. Cells adopt spherical or quasi-spherical shapes because this geometry minimizes surface area relative to volume, optimizing energy efficiency for processes like nutrient exchange and waste removal. This form follows the principle of least action, a cornerstone of physical laws that governs the behavior of systems seeking equilibrium.

Bekenstein (2003) draws a parallel between the spherical shape of cells and the geometry of black holes, which are also optimized to maximize entropy while minimizing surface area. In black holes, the surface area of the event horizon corresponds to the amount of information the black hole can contain. Similarly, the surface of a cell regulates the exchange of energy and information between the internal and external environments. This analogy underscores the universality of mathematical principles, linking cellular geometry to the broader structure of the cosmos.

Cells also embody the recursive and iterative dynamics seen in mathematical systems. The division and differentiation of cells during development mirror fractal patterns, where each iteration builds upon the previous structure to create complexity. These processes demonstrate how life organizes itself through principles of mathematical symmetry and efficiency, reflecting the order present in the universe.

### *Feedback Loops: Life as a Self-Excited Circuit*

John Wheeler's concept of the "self-excited circuit" (1994) positions life as a feedback loop between the observer and the cosmos. In Wheeler's view, the universe is not a static entity but a participatory system where observation plays a critical role in shaping reality. This idea aligns with quantum mechanics, where the act of measurement influences the state of a system.

Life itself functions as a self-excited circuit, with feedback loops driving adaptation, growth, and evolution. At the cellular level, feedback mechanisms regulate processes such as gene expression, metabolic pathways, and homeostasis. These loops create dynamic systems that respond to environmental stimuli, ensuring survival and reproduction. The self-referential nature of these processes echoes Wheeler's notion of a universe that observes itself through its components.

On a larger scale, the relationship between life and the cosmos can be seen as a macro-level feedback loop. Organisms evolve in response to cosmic phenomena, such as solar radiation and gravitational forces, while contributing to the planetary systems that sustain them. This interplay highlights the interconnectedness of life and the universe, with each influencing and shaping the other.

### *Life as the Embodiment of Cosmic Mathematics*

Framing life as a mathematical simulacrum emphasizes its role as an emergent property of universal laws. Tegmark (2015) argues that the cosmos itself is a mathematical structure, with its patterns, symmetries, and dynamics reflected in the behavior of biological systems. The spherical geometry of cells, the fractal organization of physiological networks, and the feedback loops that drive adaptation all demonstrate how life embodies these principles.

Life's ability to self-organize, adapt, and evolve arises from the same mathematical rules that govern the cosmos. Whether through the geometry of a cell or the recursive processes of evolution, biological systems reveal the deep connections between the microcosmic and the macrocosmic. By understanding life as a representation of cosmic mathematics, we gain a greater appreciation for its place within the broader tapestry of existence.

## Homologies Between Mathematics and Cellular Evolution

The parallels between mathematical structures and biological processes provide profound insights into the principles driving cellular evolution. Mathematics, as the universal language of patterns and relationships, offers a framework for understanding the organization, adaptation, and development of life. This section examines key homologies between mathematics and cellular processes, focusing on knot theory, the Fibonacci sequence, and Peter Rowlands' Rewrite Mathematics to explore how these frameworks illuminate the dynamics of evolution.

## *Knot Theory as a Metaphor for Cellular Division*

Knot theory, a branch of topology that studies the properties of intertwined loops and their transformations, serves as a compelling metaphor for cellular division. Kauffman (2006) highlights how the untangling and reformation of knots during mitosis and meiosis mirror the complex processes of chromosome segregation and cellular reorganization.

During mitosis, chromosomes condense into tightly coiled structures, resembling knots. These structures must untangle and align at the metaphase plate to ensure accurate segregation into daughter cells. Similarly, in meiosis, the process of homologous recombination involves the exchange of genetic material between paired chromosomes, creating temporary "knots" that are resolved to facilitate genetic diversity. The dynamics of these processes reflect the mathematical principles of knot theory, where the untying and retangling of loops create new configurations.

This metaphor emphasizes the precision and adaptability of cellular division, showing how biological systems navigate complexity through iterative transformations. Knot theory provides a lens for understanding how cellular systems maintain order while introducing variability, a balance essential for evolution.

## *The Fibonacci Sequence and the Golden Ratio in Cellular Evolution*

The Fibonacci sequence and the Golden Ratio, iconic mathematical constructs, appear ubiquitously in nature and play a significant role in cellular evolution. Livio (2002) describes these patterns as homologs for growth and adaptation, providing a mathematical framework for understanding physiological responses to environmental changes.

The Fibonacci sequence, where each term is the sum of the two preceding ones, manifests in biological structures such as phyllotaxis (leaf arrangement) and branching patterns in trees. These configurations optimize resource acquisition, demonstrating how mathematical efficiency governs cellular and organismal design. Similarly, the Golden Ratio (ϕ ≈ 1.618\phi \approx 1.618ϕ ≈ 1.618) appears in the proportions of biological structures, from DNA's helical dimensions to the curvature of seashells, indicating its evolutionary significance in maximizing functional harmony and aesthetic balance.

In cellular systems, these mathematical principles are evident in processes like cell division and morphogenesis. The distribution of cells in tissues often follows Fibonacci-like arrangements, optimizing spatial efficiency and signaling pathways. By aligning biological patterns with mathematical constructs, the Fibonacci sequence and Golden Ratio highlight the universality of evolutionary principles across scales.

### *Rewrite Mathematics and Attractor Dynamics in Evolution*

Peter Rowlands' Rewrite Mathematics (2015) offers a novel framework for understanding the attractor dynamics of cellular evolution. This approach models evolution as a series of recursive transformations, where systems are continually "rewritten" in response to environmental pressures.

Attractor dynamics describe how biological systems stabilize around specific configurations or behaviors, representing optimal solutions to adaptive challenges. For example, cellular energy flows, mediated by mitochondrial activity, exhibit attractor-like behavior, ensuring efficient metabolism despite fluctuations in environmental conditions. Rewrite Mathematics captures these dynamics, illustrating how cellular systems navigate evolutionary landscapes by iteratively refining their structures and functions.

Rowlands' framework also sheds light on the integration of genetic, epigenetic, and environmental factors in evolution. By viewing these interactions through a mathematical lens, we gain insights into the mechanisms driving complexity and innovation in cellular systems. This perspective positions cellular evolution as a dynamic process governed by universal mathematical rules.

### *Conclusion*

Homologies between mathematical structures and cellular evolution reveal the deep interconnectedness of life and the cosmos. Knot theory provides a metaphor for the intricate processes of cellular division, highlighting the precision and adaptability of biological systems. The Fibonacci sequence and the Golden Ratio offer mathematical homologs for growth and adaptation, demonstrating their role in optimizing physiological designs. Peter Rowlands' Rewrite Mathematics frames evolution as a dynamic interplay of attractor states, capturing the iterative transformations that drive complexity. Together, these perspectives underscore the universality of mathematical principles in shaping the patterns and processes of life.

## The Cell as a Holographic Event Horizon

The cell, the fundamental unit of life, can be conceptualized as a holographic system that reflects the principles governing cosmic structures. Its boundary—the cell membrane—not only defines its physical limits but also embodies dynamic interactions between energy, matter, and information. This perspective aligns with the ideas of Bekenstein (2003), who proposed that surface area is a measure of information content, and Schrödinger (1944), who framed life as a battle against entropy. Torday's (2021) notion of the cell as a Möbius Strip further underscores the duality between the local and nonlocal properties of cellular processes. Together, these concepts position the cell as a microcosmic event horizon, integrating life with universal laws.

## *The Cell Membrane as an Event Horizon*

Bekenstein's (2003) insights into black holes and entropy provide a powerful analogy for understanding the cell membrane as an event horizon. In Black Holes, the event horizon represents the boundary beyond which no information or matter can escape. The surface area of this boundary correlates with the Black Hole's entropy and informational content. Similarly, the cell membrane acts as a dynamic interface, regulating the exchange of energy, nutrients, and signals between the internal and external environments.

The membrane's structure—a bilayer of lipids embedded with proteins—optimizes the cell's ability to maintain homeostasis while adapting to external pressures. This adaptability mirrors the way Black Holes interact with their surroundings, absorbing and processing information to achieve equilibrium. Just as the event horizon encapsulates the fundamental properties of a black hole, the cell membrane encodes the information necessary for life, balancing entropy and order.

The holographic nature of the cell is evident in its capacity to store and process information. Genetic material, epigenetic modifications, and signaling pathways all contribute to a cellular "information network" that mirrors the principles of holography. This analogy reveals the interconnectedness of cellular processes and cosmic phenomena, demonstrating how life adheres to universal principles of energy and information dynamics.

## *The Cell as a Möbius Strip*

Torday (2021) proposed that the cell can be conceptualized as a Möbius Strip—a surface with only one side and one edge—symbolizing its dual nature. This duality reflects the interplay between local and nonlocal properties, where cellular processes operate within defined boundaries while remaining interconnected with the broader system.

For example, ion channels and transport proteins within the membrane facilitate the selective exchange of molecules, enabling the cell to interact with its environment. These local processes are influenced by nonlocal factors, such as electromagnetic fields and Quantum Coherence, which guide cellular behavior at a systemic level. The Möbius Strip metaphor encapsulates this dynamic, illustrating how the cell functions as both an independent unit and an integral part of the larger biological and cosmic networks.

The Möbius Strip also underscores the cell's ability to integrate opposing forces, such as entropy and negentropy. This balance is crucial for sustaining life, allowing the cell to adapt to changing conditions while preserving its core functionality.

## *Negentropy and the Foundations of Life*

Schrödinger's (1944) concept of negentropy, or negative entropy, highlights the cell's ability to resist disorder by maintaining homeostasis. Unlike inanimate systems, which tend toward entropy, living organisms actively harness energy to sustain order and complexity. The cell achieves this by continually exchanging matter and energy with its environment, using metabolic pathways to convert resources into usable forms.

Negentropy is evident in processes such as ATP production, where energy from nutrients is transformed into a form that powers cellular functions. This energy is used to maintain structural integrity, repair damage, and drive growth and reproduction. By countering the effects of entropy, the cell ensures its survival and adaptability, embodying Schrödinger's vision of life as an "island of order" in a sea of chaos.

The holographic nature of the cell extends to its ability to encode and transmit information. Through genetic replication and signaling networks, cells pass on the instructions necessary for maintaining homeostasis and responding to environmental challenges. This dynamic interplay of energy, matter, and information positions the cell as a fundamental unit of life that mirrors the principles of cosmic order.

## *Conclusion*

The cell, as a holographic event horizon, integrates principles from mathematics, physics, and biology to sustain life. Its membrane serves as a dynamic boundary that regulates information and energy flow, akin to the event horizon of a black hole. The Möbius Strip metaphor captures the cell's duality, balancing local and nonlocal processes to achieve systemic integration. Schrödinger's concept of negentropy underscores the cell's role in maintaining order against the forces of entropy. Together, these perspectives reveal the cell as a microcosmic reflection of the universe, demonstrating the profound interconnectedness of life and the cosmos.

## Symbiogenesis and the Passive Assimilation of Mathematics

Symbiogenesis, the merging of distinct biological entities to form new adaptive systems, is a powerful driver of evolution. This process illustrates how organisms integrate environmental challenges into their physiology, passively incorporating mathematical principles to optimize survival. By linking cellular processes to universal laws, symbiogenesis demonstrates how life is shaped by both its immediate surroundings and the broader cosmos. This section explores the role of symbiogenesis, epigenetic inheritance, and directed mutation in framing evolution as an adaptive and mathematically informed process.

### *Symbiogenesis: Integrating Mathematics into Physiology*

The theory of symbiogenesis, as articulated by Chapman and Margulis (1998), posits that evolutionary innovation often arises from the merging of different organisms. This process allows for the assimilation of external challenges into physiological functions, where mathematical principles, such as efficiency and optimization, are passively incorporated into biology.

A classic example of symbiogenesis is the integration of mitochondria into ancestral eukaryotic cells. This symbiotic relationship, established billions of years ago, allowed host cells to harness energy more efficiently through oxidative phosphorylation. The process reflects mathematical optimization, as mitochondria maximize energy output relative to resource input. Such partnerships illustrate how symbiogenesis embeds universal principles, such as energy efficiency, into the fabric of life.

Symbiogenesis also facilitates the incorporation of environmental challenges into biological systems. For instance, cyanobacteria became chloroplasts within plant cells, enabling photosynthesis—a process that adheres to principles of energy conversion and conservation. These examples highlight how symbiogenesis passively integrates mathematical frameworks into the evolution of life, ensuring that organisms are better equipped to navigate their environments.

### *Epigenetic Inheritance and the Transmission of Adaptations*

Epigenetic inheritance provides a mechanism for transmitting environmental adaptations across generations, linking cellular processes to cosmic mathematics. Ben Maamar et al. (2021) describe how epigenetic modifications, such as DNA methylation and histone acetylation, regulate gene expression in response to external stimuli. These changes can be passed on to offspring, allowing organisms to "remember" and adapt to their environments over time.

Epigenetic mechanisms reflect mathematical principles by dynamically adjusting gene expression to maintain homeostasis. For example, during periods of environmental stress, specific genes may be silenced or activated to optimize resource use and energy balance. This regulation mirrors feedback loops in mathematical systems, where inputs are continuously adjusted to achieve stability.

Moreover, epigenetics bridges the gap between local cellular processes and universal dynamics. By encoding information about environmental conditions into the genome, epigenetic inheritance aligns biological adaptation with broader patterns of energy flow and optimization. This perspective positions epigenetics as a mediator between the microcosmic and macrocosmic, illustrating the interconnectedness of life and the universe.

## *Directed Mutation: Evidence for Nonrandom Evolution*

Recent evidence suggests that evolution is not entirely random, but can exhibit directed responses to environmental stress. Melamed et al. (2022) highlight how organisms facing specific challenges, such as nutrient scarcity or temperature fluctuations, can induce mutations that improve their chances of survival. This directed mutation challenges the traditional view of evolution as a purely stochastic process, emphasizing its adaptive and purposeful nature.

Directed mutation aligns with mathematical principles by demonstrating how biological systems optimize their responses to external pressures. For example, bacteria exposed to antibiotics may develop resistance through mutations that enhance their survival. These mutations are not arbitrary but are often targeted to specific genes involved in drug metabolism or transport. Such adaptability reflects the iterative and feedback-driven nature of evolutionary processes, where organisms refine their responses to achieve optimal outcomes.

The concept of directed mutation also underscores the role of environmental feedback in shaping evolution. Just as symbiogenesis incorporates external elements into physiology, directed mutation integrates environmental signals into the genetic code, creating a dynamic interplay between biology and the cosmos. This interplay reveals evolution as a process guided by universal laws of adaptation and efficiency.

## *Conclusion*

Symbiogenesis, epigenetic inheritance, and directed mutation illustrate how evolution passively assimilates mathematical principles into biological systems. Through symbiotic partnerships, organisms integrate environmental challenges into their physiology, optimizing energy use and resource management. Epigenetic mechanisms transmit these adaptations across generations, linking cellular processes to universal patterns of homeostasis and optimization. Directed mutation further highlights the nonrandom nature of evolution, demonstrating how organisms respond purposefully to external pressures. Together, these processes position evolution as a dynamic interplay between biology and cosmic mathematics, reflecting the interconnectedness of life and the universe.

## Artistic and Philosophical Reflections of Cosmic Mathematics

The profound connection between life and cosmic mathematics extends beyond biology and physics into the realms of art, music, and philosophy. Artistic creations and peak human experiences, such as Near-Death Experiences (NDEs) and meditation, reflect the mathematical patterns that underpin both cellular processes and the cosmos. This section explores these interdisciplinary connections, emphasizing the interplay between science and the humanities. Drawing on the works of Fibonacci-inspired compositions, physiological alignments with cosmic principles, and C.P. Snow's "Two Cultures" concept, it presents a holistic framework for understanding the relationship between mathematics, consciousness, and the human experience.

### *Mathematics in Art, Music, and Literature*

The Fibonacci sequence and the Golden Ratio are among the most compelling examples of mathematics influencing art and music. These principles, found throughout nature, have inspired compositions, visual art, and literary structures that echo the harmony and balance of the cosmos. Saunders (2022) highlights how composers like Béla Bartók and visual artists such as Salvador Dalí deliberately employed Fibonacci proportions to evoke a sense of natural order and aesthetic resonance.

For example, Bartók's *Music for Strings, Percussion, and Celesta* incorporates Fibonacci-inspired structures in its phrasing and rhythm, creating a composition that mirrors the recursive and harmonious dynamics of the natural world. Similarly, in visual art, the Golden Ratio appears in the arrangement of elements to create balance and proportion, reflecting the same optimization principles seen in cellular evolution.

In literature, Fibonacci patterns are used to structure narratives, with recursive themes and proportions enhancing the storytelling. These creative expressions demonstrate how humans instinctively align their artistic endeavors with mathematical principles, reflecting the universal patterns that govern both life and the cosmos.

## Physiological Alignments and Cosmic Mathematics

Human experiences such as Near-Death Experiences (NDEs), Peak Experiences, and meditation provide physiological insights into the alignment between consciousness and cosmic mathematics. Torday (2020) suggests that such experiences represent states where the body's internal rhythms synchronize with universal principles, creating a profound sense of unity and transcendence.

NDEs, often described as encounters with ineffable light and harmony, may reflect the brain's alignment with the energy flows and recursive patterns of the cosmos. Similarly, Peak Experiences, characterized by moments of heightened awareness and interconnectedness, mirror the feedback loops seen in cellular processes, where balance and efficiency optimize function.

Meditation offers another pathway for aligning physiology with cosmic mathematics. Practices such as mindfulness and transcendental meditation promote homeostasis by regulating breathing, heart rate, and neural activity. This physiologic coherence resonates with mathematical principles, such as symmetry and harmonic oscillation, fostering a sense of unity between the individual and the universe.

These experiences highlight the interconnectedness of life, consciousness, and cosmic order, suggesting that human awareness is intrinsically linked to the mathematical fabric of existence.

## The "Two Cultures" Revisited: Bridging Science and the Humanities

C.P. Snow's *The Two Cultures* (1959) identified a divide between the sciences and the humanities, emphasizing the need for interdisciplinary integration. Cellular and cosmic mathematics offer a unifying framework that bridges this gap, demonstrating how scientific principles inform artistic expression and philosophical inquiry.

For example, the study of cellular processes reveals the mathematical underpinnings of life, while the humanities explore the meaning and aesthetics of these patterns. The recursive dynamics of cellular evolution can inspire narrative structures in literature, just as the Golden Ratio shapes artistic compositions. This integration fosters a holistic understanding of the human experience, where scientific discovery and creative expression enrich one another.

Snow's call for interdisciplinary dialogue finds new relevance in the context of cellular and cosmic mathematics. By framing life as an emergent property of universal laws, this perspective unites the analytical rigor of science with the intuitive insights of art and philosophy, offering a comprehensive view of existence.

## *Conclusion*

Artistic and philosophical reflections of cosmic mathematics reveal the deep connections between life, consciousness, and the universe. Fibonacci-inspired compositions, NDEs, and meditation exemplify how mathematical principles influence human creativity and awareness. Revisiting Snow's "Two Cultures" concept, this section proposes a holistic integration of science and the humanities, positioning cellular and cosmic mathematics as a bridge between disciplines. Through this interdisciplinary lens, we gain a richer understanding of life as both a biological phenomenon and an artistic expression of the cosmos.

## References


Bekenstein, J. D. (2003). Information in the holographic universe. *Scientific American, 289*, 58–65.

Ben Maamar, M., Nilsson, E. E., & Skinner, M. K. (2021). Epigenetic transgenerational inheritance. *Biology of Reproduction, 105*, 570–592.

Chapman, M. J., & Margulis, L. (1998). Morphogenesis by symbiogenesis. *International Microbiology, 1*, 319–326.

Kauffman, L. (2006). *Formal knot theory*. Dover Publications.

Livio, M. (2002). *The golden ratio*. Broadway Books.

Melamed, D., et al. (2022). De novo mutation rates and adaptation. *Genome Research, 14*, 488–498.

Rowlands, P. (2015). *The foundations of physical law*. World Scientific.

Saunders, G. (2022). *A swim in a pond in the rain*. Random House.

Schrödinger, E. (1944). *What is life?* Macmillan.

Snow, C. P. (1959). *The two cultures*. Cambridge University Press.

Tegmark, M. (2015). *Our mathematical universe*. Vintage.

Torday, J. S. (2020). Consciousness, redux. *Medical Hypotheses, 140*, 109674.

Torday, J. S. (2021). Life is a Möbius strip. *Progress in Biophysics and Molecular Biology, 167*, 41–45.

Wheeler, J. (1994). *At home in the universe*. American Institute of Physics.

# Chapter 7
# The Geometry of Consciousness and Evolution

Geometry, the study of shapes and their relationships, transcends its mathematical origins to offer profound insights into the processes that define life and consciousness. From the intricate patterns of cellular structures to the expansive dynamics of cosmic evolution, geometry provides a universal framework for understanding how complexity arises from simplicity. This chapter explores the interplay between geometric principles and the evolutionary processes that shape life, positioning geometry as the bridge between the physical and the metaphysical.

Consciousness and evolution, though often studied independently, are deeply intertwined through geometric patterns and dynamics. The self-organizing principles of life, such as symmetry, fractals, and spirals, are not merely aesthetic, but serve as functional mechanisms for optimizing energy use, facilitating adaptation, and sustaining life. These same principles manifest in the neural networks of the brain, where geometric arrangements underpin cognition and self-awareness, and in the cosmic lattice of the universe, where structure and motion emerge from fundamental laws.

This chapter begins by examining the role of symmetry and asymmetry in evolutionary processes, demonstrating how these geometric properties drive diversity and innovation in biological systems. It then delves into the fractal nature of physiological networks, highlighting how recursive patterns optimize efficiency and connectivity. The chapter also explores the influence of geometry on consciousness, linking the spatial and temporal dimensions of neural activity to the emergent properties of awareness. Finally, it reflects on the broader implications of geometry as a unifying principle that connects the microcosm of cellular life with the macrocosm of the universe.

Through the lens of geometry, this chapter seeks to illuminate the profound interconnectedness of evolution and consciousness, offering a deeper understanding of their shared foundations in the universal patterns that govern existence.



J. S. Torday, R. G. Sacco, *Conscious Life*,
https://doi.org/10.1007/978-3-031-80500-4_7

## The Origins of Topological Consciousness

The origins of consciousness can be traced to the fundamental processes that govern cellular life. The hypothesis of topological consciousness posits that awareness arises from the cumulative memory of evolutionary adaptations encoded within cellular structures and functions. Topology, the study of spatial properties preserved under deformation, offers a powerful framework for understanding how life organizes itself to achieve equilibrium and adaptation. This section explores the formation of micelles as the primordial boundary for life, the balance between entropy and homeostasis, and the role of knot theory as a metaphor for cellular processes.

### *Micelle Formation: The Primordial Surface for Topological Consciousness*

Micelles, spherical arrangements of lipid molecules, represent the earliest known biological boundaries separating internal and external environments. These self-assembling structures, driven by hydrophobic and hydrophilic interactions, created the first "surface" for life—a critical step in the emergence of topological consciousness (Deamer, 2017).

The micelle's boundary was more than a physical separation; it established the framework for communication and energy exchange with the environment. This primitive topology allowed life to maintain internal order while interacting with external chaos, laying the groundwork for more complex cellular structures. Torday and Rehan (2012) emphasize that this initial differentiation between "inside" and "outside" formed the template for cellular self-awareness, where the ability to distinguish self from nonself became a fundamental property of life.

In this context, the micelle can be seen as the first topological surface encoding environmental memory. The lipid bilayer's ability to adapt and reorganize in response to external stimuli mirrored the principles of evolution, creating a dynamic system that preserved and transmitted information across generations.

### *Negative Entropy and Functional Equilibrium*

The cell's ability to maintain homeostasis against the backdrop of environmental entropy is central to its survival and adaptability. Schrödinger's (1944) concept of "negative entropy" describes how living systems draw energy from their surroundings to counteract the natural tendency toward disorder. Torday and Miller (2017) expand on this idea, suggesting that cellular processes achieve a state of ambiguous but functional equilibrium, where opposing forces are balanced to sustain life.

For example, ion gradients across the cell membrane represent a form of topological tension, where energy is stored and utilized to drive essential functions such as nutrient transport and signal transduction. This dynamic equilibrium reflects the cell's capacity to integrate external entropy into its internal order, creating a stable yet adaptable system.

This balance between entropy and homeostasis is a hallmark of topological consciousness. The cell's ability to navigate these opposing forces demonstrates how evolutionary memory is encoded not only in genetic material but also in the spatial organization of cellular processes.

## *Knot Theory: A Metaphor for Cellular Processes*

Knot theory, a branch of mathematics that examines the properties of intertwined loops, provides a compelling metaphor for understanding cellular structures and their resilience. Kauffman (2006) describes mathematical knots as configurations that maintain their integrity under deformation, a property that parallels the behavior of physiologic systems.

In cellular biology, "knots" can be seen in the coiling of DNA, the folding of proteins, and the organization of the cytoskeleton. These structures must remain functional despite the dynamic forces acting upon them. For instance, during mitosis, chromosomes condense into tightly coiled loops to ensure accurate segregation. This process reflects the principles of knot theory, where complex arrangements adapt to external pressures without losing their fundamental properties.

The application of knot theory to cellular processes highlights the topological nature of evolution. Just as mathematical knots encode spatial information, physiological "knots" preserve the structural and functional integrity of cells, enabling them to adapt and evolve in response to environmental challenges.

## *Conclusion*

The origins of topological consciousness lie in the cellular processes that encode evolutionary memory and maintain functional equilibrium. Micelle formation provided the first boundary for life, creating a surface for self-awareness and environmental interaction. The balance between negative entropy and external chaos highlights the cell's ability to sustain order in dynamic environments. Knot theory offers a metaphor for the resilience and adaptability of cellular structures, emphasizing the role of topology in preserving life's complexity. Together, these principles illustrate how consciousness emerges from the spatial and functional properties of living systems.

## Symbiogenesis and the Evolutionary Dynamics of Knots

Symbiogenesis, the evolutionary process where distinct organisms merge to form a new biological entity, has profoundly shaped life's trajectory. This integration is not just a biological phenomenon but also a topological transformation, akin to untying and retying knots to form more complex configurations. Through symbiogenesis, life internalizes environmental challenges, generating innovative homeostatic mechanisms and evolutionary pathways. This section explores the role of symbiogenesis in cellular evolution, its topological implications, and its contribution to multicellular complexity and speciation.

### *Symbiogenesis: Internalizing Environmental Threats*

Symbiogenesis is a powerful evolutionary strategy that integrates environmental threats into physiology, creating novel adaptive solutions. Sagan (1967) highlighted this process as central to the origin of eukaryotic cells, where free-living prokaryotes became mitochondria and chloroplasts within host cells. These symbiotic relationships transformed environmental challenges, such as the need for efficient energy production, into physiological advantages.

Ben Maamar et al. (2021) expanded on this concept, showing that symbiogenesis extends beyond organelle formation to encompass epigenetic modifications. Cells can incorporate external signals into their regulatory networks, creating lasting changes that are transmitted across generations. For instance, epigenetic inheritance allows organisms to adapt rapidly to environmental pressures without waiting for genetic mutations, reflecting the dynamic interplay between internal processes and external forces.

This adaptive mechanism parallels the untying and retying of knots in topology. Just as a knot's configuration can be altered to stabilize under tension, cellular symbiosis reshapes internal systems to meet external demands. By integrating external elements, symbiogenesis represents a topological transformation that drives evolutionary innovation.

### *Homology Between Cellular Evolution and Knot Formation*

The evolutionary dynamics of symbiogenesis can be understood through the lens of knot theory, which examines the properties of intertwined loops. Torday (2019) proposed that cellular evolution mirrors the formation and resolution of physiological knots. For example, during aging and senescence, cellular systems encounter "knots" in the form of accumulated damage and dysregulated signaling. Untying these knots involves processes like autophagy and DNA repair, which restore cellular function and maintain homeostasis.

Knot formation also applies to developmental biology, where the organization of tissues and organs relies on precise topological arrangements. The folding and looping of DNA, the branching of vasculature, and the spatial alignment of cells during morphogenesis all exhibit knot-like properties that ensure structural integrity and functionality. These dynamic transformations illustrate how life evolves by manipulating its internal topology to adapt and thrive.

The untying of knots during cellular stress and aging exemplifies the resilience of biological systems. By resolving topological challenges, cells preserve their ability to function and adapt, reflecting the continuous interplay between form and function in evolution.

## *Knot-Like Interactions and Evolutionary Complexity*

As life transitioned from unicellular to multicellular organisms, the interactions between cells became increasingly knot-like, characterized by intricate connections and dependencies. Demayo et al. (2022) emphasized that these interactions are central to multicellular integration and speciation. In multicellular systems, cells communicate and coordinate their activities through complex signaling networks, resembling the interwoven structure of knots.

These interactions enable organisms to achieve higher levels of complexity and specialization. For example, the evolution of nervous systems required the precise integration of neuronal connections, forming a topological network that supports information processing and adaptive behavior. Similarly, the development of vascular systems relied on branching patterns that optimize resource distribution, reflecting the geometric and topological principles underlying biological organization.

The relationship between cellular knots and evolutionary complexity highlights the importance of topology in speciation. By reshaping and refining their internal configurations, organisms can adapt to new environments, exploit novel niches, and diversify into distinct species. This dynamic underscores the role of symbiogenesis as a topological driver of evolutionary innovation.

## *Conclusion*

Symbiogenesis and the dynamics of knots illustrate the profound interconnectedness between biology and topology. By internalizing environmental challenges, symbiogenesis transforms external pressures into adaptive advantages, mirroring the reconfiguration of knots under tension. The homology between cellular evolution and knot formation reveals the resilience and adaptability of life, while knot-like interactions in multicellular systems drive complexity and speciation. Together, these processes demonstrate how topology shapes the evolutionary trajectory of life, offering a unifying framework for understanding its intricate dynamics.

## Cellular Memory and Evolutionary Exaptations

Cellular memory, the ability of biological systems to retain and utilize information about past environmental interactions, plays a foundational role in evolutionary adaptability. This memory is embedded in molecular structures, cellular processes, and genetic mechanisms, enabling organisms to respond to challenges and innovate solutions over generations. Evolutionary exaptations—traits co-opted for new functions—arise from this reservoir of cellular memory. This section explores the molecular memory of lipids and micelles, the concept of cellular hysteresis, and the role of epigenetic marks in transferring environmental data to subsequent generations.

### *Lipids and Micelles: Molecular Memory as Evolutionary Foundations*

Lipids and micelles, fundamental components of cellular structures, exhibit molecular memory that serves as the basis for evolutionary adaptability. Gould and Vrba (1982) introduced the concept of exaptations, where existing traits are repurposed for new functions. Micelles, the precursors to cellular membranes, exemplify this principle by adapting their structural properties to environmental conditions.

The self-assembling nature of micelles, driven by the hydrophobic and hydrophilic interactions of lipids, creates a dynamic boundary that retains information about external pressures. For instance, micelles adjust their size, shape, and permeability in response to changes in temperature, pH, or salinity. These adjustments encode environmental data into their structure, forming a type of molecular memory that guides future adaptations.

This memory becomes crucial during evolutionary transitions, such as the development of more complex cellular membranes in eukaryotes. The flexibility and responsiveness of micelles provided a foundation for the iterative refinement of lipid bilayers, enabling the emergence of cellular systems that could sustain life in diverse environments. By preserving and reinterpreting molecular information, lipids and micelles contribute to the adaptive toolkit of evolution.

### *Cellular Hysteresis and Environmental Adaptability*

Cellular hysteresis, the phenomenon where a system's response to current conditions depends on its past states, serves as a precursor to the genetic memory stored in DNA. Torday and Miller (2017) describe how cellular systems "remember" environmental stimuli through changes in protein activity, signaling pathways, and membrane dynamics. This memory allows cells to anticipate and prepare for recurring challenges, enhancing their adaptability.

For example, stress-induced modifications in ion channel activity or metabolic enzyme expression can persist even after the stressor is removed. These changes create a feedback loop where past experiences influence future responses, mirroring the principles of hysteresis observed in physical systems. Over time, these transient modifications may become encoded in the genome through genetic mutations or epigenetic marks, translating cellular memory into evolutionary innovation.

Hysteresis is particularly evident in the development of complex traits, such as thermoregulation or immune responses. By retaining information about environmental fluctuations, cells can fine-tune their functions to optimize survival. This iterative process, where cellular memory informs and shapes evolutionary trajectories, underscores the dynamic interplay between biology and environment.

## *Epigenetic Marks and the Transfer of Environmental Data*

Epigenetic mechanisms provide a bridge between cellular memory and genetic evolution by encoding environmental data into heritable marks. Ben Maamar et al. (2021) highlight how modifications such as DNA methylation and histone acetylation regulate gene expression in response to external stimuli. These epigenetic marks can persist across generations, enabling organisms to "inherit" adaptive responses to environmental challenges.

For example, exposure to nutritional stress or toxins during one generation can lead to epigenetic changes that influence the health and behavior of subsequent generations. These modifications reflect an iterative cycle of cellular evolution, where environmental inputs are integrated into the genome without altering the underlying DNA sequence. This process allows organisms to rapidly adapt to changing conditions while preserving the flexibility to evolve further.

Epigenetic inheritance exemplifies the principle of exaptation, as traits initially acquired in response to specific challenges may later be repurposed for new functions. This dynamic underscores the role of cellular memory as a driver of evolutionary innovation, where information from the past informs the adaptations of the future.

## *Conclusion*

Cellular memory serves as the foundation for evolutionary adaptability, enabling organisms to respond to environmental challenges and innovate solutions over generations. Lipids and micelles exhibit molecular memory that informs structural and functional adaptations, laying the groundwork for evolutionary processes like exaptation. Cellular hysteresis demonstrates how past experiences shape future responses, creating a feedback loop between environment and biology. Epigenetic marks provide a mechanism for transferring environmental data across generations, integrating cellular memory into the evolutionary narrative. Together, these processes highlight the central role of cellular memory in shaping the trajectory of life.

## The Metaphysics of Knots and Cellular Topology

The interplay between cellular topology and consciousness transcends biology, reaching into the realms of metaphysics and philosophy. Cellular processes, with their intricate topological transformations, reflect deeper connections between the explicate and implicate orders described by Bohm (1980). These transformations offer insights into the emergence of consciousness and its dual subjective and objective dimensions. This section explores how cellular topology bridges these realms, linking biological processes to metaphysical interpretations of knots in art, science, and consciousness.

### *Cellular Topology and Bohm's Explicate and Implicate Orders*

David Bohm's (1980) theory of the explicate and implicate orders provides a philosophical framework for understanding cellular topology. In Bohm's view, the explicate order represents the manifest, observable world, while the implicate order describes the underlying, interconnected reality from which the explicate emerges. Cellular processes, particularly topological transformations, bridge these two realms, encoding the complexity of life in dynamic yet coherent patterns.

The folding and unfolding of cellular membranes, akin to the behavior of a Möbius strip, exemplify this bridging. The cell's topology allows for the simultaneous containment and interaction of internal and external environments, creating a system that is both discrete and interconnected. These transformations reflect Bohm's idea that the explicate and implicate orders are not separate but dynamically intertwined.

For example, cellular signaling pathways involve both localized biochemical reactions and global systemic responses, illustrating how individual processes contribute to the organism's holistic function. These interactions resonate with Bohm's implicate order, where every part contains the whole. Cellular topology, therefore, provides a tangible model for exploring the emergent nature of consciousness as an interplay between the microcosmic and macrocosmic.

## *Lateralization and Quantum Topology in Consciousness*

The lateralization of brain function—where distinct hemispheres specialize in different cognitive tasks—offers a biological example of topological resolution, linking quantum mechanics to perception and cognition. Torday (2019) suggests that lateralization represents a unifying topology in biology, resolving quantum uncertainties into coherent subjective experiences.

Quantum mechanics, with its principles of superposition and entanglement, mirrors the dualities inherent in consciousness, such as the separation between self and other or subject and object. Lateralization transforms these dualities into functional asymmetries that enable advanced cognitive processes, such as language and abstract reasoning. This process can be seen as a "topological collapse," where the potentialities of quantum states resolve into the deterministic structures of perception and thought.

By framing lateralization as a topological phenomenon, we can understand consciousness as an emergent property of cellular and neural interactions. This perspective links the quantum and biological realms, illustrating how the metaphysics of topology informs the experience of consciousness.

## *Knots as Metaphors for Subjective and Objective Dimensions*

Knots have long been used as metaphors in art and science to explore the tension between order and complexity. René Magritte's (1937) painting *The Treachery of Images* offers a visual representation of this tension, where the depicted object—a pipe—highlights the gap between representation and reality. Similarly, in science, knot theory provides a mathematical framework for understanding intertwined structures and their transformations.

In cellular biology, knots represent the intricate configurations of DNA, protein folding, and cytoskeletal dynamics, all of which maintain the integrity and functionality of life. These biological knots reconcile the subjective and objective dimensions of existence, where cellular processes generate both individual experiences and shared realities.

Bohm (1980) posits that knots symbolize the connection between the implicate and explicate orders, representing the dynamic interplay between hidden potentialities and manifest phenomena. Cellular knots, with their dual role in preserving genetic information and enabling adaptation, reflect this metaphysical interpretation, bridging the tangible and the abstract.

### *Conclusion*

The metaphysics of knots and cellular topology reveals the profound connections between biology, philosophy, and consciousness. Cellular processes, as topological transformations, bridge Bohm's explicate and implicate orders, linking the manifest and interconnected dimensions of reality. Brain lateralization illustrates how quantum mechanics resolves into the topology of perception and cognition, creating a unified framework for understanding consciousness. Knots, as metaphors for the subjective and objective dimensions of existence, underscore the interplay between complexity and order in cellular and cosmic systems. Together, these perspectives position cellular topology as a foundational principle for exploring the emergence of life and consciousness.

## Evolutionary Insights from Topological Models

The integration of topology into evolutionary biology provides a powerful framework for understanding the emergence and organization of life. Cellular topology, with its emphasis on spatial and structural relationships, informs the development of complex systems while reflecting broader universal principles. This section synthesizes insights from cellular topology, quantum mechanics, and evolutionary theory into a unified model of life and consciousness, highlighting the phylogenetic origins of complex systems, algorithmic approaches to evolution, and the cosmic implications of topological organization.

### *Cellular Topology and the Development of Complex Systems*

The evolution of complex organ systems, such as the lung, heart, and kidney, can be understood through the lens of cellular topology. These systems, which enable specialized functions like gas exchange, circulation, and filtration, evolved from simpler ancestral structures by adapting to environmental pressures and optimizing spatial organization.

Demayo et al. (2022) emphasize how the phylogenetic origins of these systems are rooted in cellular topology. For instance, the vertebrate lung evolved from a rudimentary swim bladder in fish, driven by the need for efficient oxygen exchange in terrestrial environments. This transition involved topological transformations, such as the folding of epithelial surfaces to increase surface area, a pattern also seen in other organs like the kidney's nephrons and the brain's cortical folds.

These examples illustrate how cellular topology facilitates the iterative refinement of form and function, enabling organisms to adapt to diverse ecological niches.

The principles of topology, such as symmetry, connectivity, and surface optimization, underpin the structural and functional evolution of these systems, demonstrating their role as universal drivers of complexity.

## *Algorithmic Approaches to Evolution*

Evolution can be viewed as an algorithmic process, where changes in cellular topology are guided by principles from quantum mechanics and biology. Torday and Miller (2017) propose that evolutionary dynamics operate as iterative feedback loops, where environmental inputs drive adaptive changes that are encoded and refined over generations. This perspective links cellular topology to algorithmic frameworks, offering a cohesive model for understanding the mechanisms of evolution.

At the quantum level, phenomena such as entanglement and superposition provide a basis for the nonlinear and probabilistic nature of evolutionary adaptations. For example, the organization of ion channels and molecular interactions within cells exhibits quantum coherence, enabling efficient energy transfer and signaling. These quantum processes influence cellular topology, shaping the structural and functional properties of life.

By integrating quantum mechanics with cellular biology, the algorithmic approach reveals evolution as a multi-scale process that connects the molecular, cellular, and systemic levels. This model highlights the iterative nature of evolution, where topological transformations serve as both drivers and outcomes of adaptive changes, creating a continuous cycle of innovation.

## *Topological Models as Universal Principles of Organization*

The principles of topology extend beyond biology, offering insights into the organization of the cosmos. Torday (2019) positions cellular topology as a microcosmic reflection of universal dynamics, where the spatial and structural properties of cells mirror the patterns seen in galaxies, black holes, and other cosmic systems.

For instance, the folding of cellular membranes can be compared to the curvature of space-time in Einstein's theory of general relativity. Both processes involve the optimization of energy and matter distribution to achieve stability and coherence. Similarly, the recursive patterns seen in fractal geometry, which describe phenomena ranging from vascular branching to galaxy formation, underscore the universality of topological organization.

By merging topology with cellular evolution, we gain a unified framework for understanding life and the cosmos. This perspective positions evolution as a process that reflects universal principles, linking the emergence of complexity in living systems to the broader dynamics of the universe.

## *Conclusion*

Topological models offer profound insights into the evolution of life and consciousness, bridging biology, physics, and cosmology. Cellular topology informs the development of complex systems like the lung, heart, and kidney, revealing the structural and functional adaptations that drive phylogenetic innovation. Algorithmic approaches to evolution highlight the iterative and multi-scale nature of adaptive changes, integrating quantum mechanics and cellular biology into a cohesive model. Finally, the universality of topological principles connects cellular evolution to the organization of the cosmos, positioning life as an emergent property of the same dynamics that govern the universe. Through this synthesis, topology emerges as a unifying framework for exploring the interconnectedness of life and existence.

## References

Ben Maamar, M., Sadler-Riggleman, I., Beck, D., & Skinner, M. K. (2021). Epigenetic transgenerational inheritance of altered sperm histone retention sites. *Environmental Epigenetics, 7*(1), dvab001.

Bohm, D. (1980). *Wholeness and the implicate order*. Routledge.

Deamer, D. (2017). The role of lipid membranes in life's origin. *Life, 7*(1), 5.

Demayo, T., Liu, Z., & Lee, R. (2022). Evolutionary dynamics of multicellular integration: A topological perspective. *Evolutionary Biology, 49*(2), 245–259.

Gould, S. J., & Vrba, E. S. (1982). Exaptation—A missing term in the science of form. *Paleobiology, 8*(1), 4–15.

Kauffman, L. H. (2006). Knot theory and biology: Untangling complexity in cellular systems. *Journal of Theoretical Biology, 241*(1), 233–243.

Magritte, R. (1937). *La trahison des images [The treachery of images]*. Los Angeles County Museum of Art.

Sagan, L. (1967). On the origin of mitosing cells. *Journal of Theoretical Biology, 14*(3), 255–274.

Schrödinger, E. (1944). *What is life? The physical aspect of the living cell*. Cambridge University Press.

Torday, J. S. (2019). The singularity of nature. *Progress in Biophysics and Molecular Biology, 142*, 23–31.

Torday, J. S., & Miller, W. B. (2017). The resolution of ambiguity as the basis for life. *Progress in Biophysics and Molecular Biology, 131*, 288–297.

Torday, J. S., & Rehan, V. K. (2012). *Evolutionary biology, cell-cell communication and complex disease*. Wiley.

# Chapter 8
# Mathematics and Cellular Evolution: Peter Rowlands' Model

The evolution of life, from its simplest forms to its most complex manifestations, is deeply intertwined with the mathematical structures that govern the universe. Mathematics not only provides the language to describe biological processes but also reveals the underlying principles that drive cellular organization, adaptation, and evolution. Among the pioneering frameworks that illuminate this connection is Peter Rowlands' *Rewrite Mathematics*, which offers profound insights into how cellular evolution aligns with the fundamental patterns of the cosmos.

In this chapter, we explore how Rowlands' mathematical model bridges the gap between biology and physics, positioning evolution as a dynamic process governed by symmetry, balance, and recursive relationships. This approach frames cellular evolution as an iterative algorithm that integrates environmental inputs, physiological processes, and genetic information into a cohesive and adaptive system.

Rowlands' model emphasizes the role of attractor dynamics in guiding evolutionary trajectories, where stability and change coexist in a delicate balance. These principles mirror the recursive structures found in mathematical phenomena such as the Fibonacci sequence and fractals, which also play key roles in cellular organization and physiological development.

The chapter begins by contextualizing Rowlands' *Rewrite Mathematics* within the broader scope of mathematical biology, examining its relevance to the principles of cellular topology and homeostasis. It then delves into how the model explains evolutionary innovation, emphasizing the interplay between randomness and determinism in cellular processes. Finally, the chapter considers the broader implications of Rowlands' framework for understanding the continuity of life, consciousness, and the cosmos.

By integrating Rowlands' mathematical insights with the dynamics of cellular evolution, this chapter seeks to uncover the universal patterns that link biology with the mathematical fabric of reality, offering a holistic perspective on the interconnectedness of life and the universe.



J. S. Torday, R. G. Sacco, *Conscious Life*,
https://doi.org/10.1007/978-3-031-80500-4_8

## The Cellular Basis of Evolutionary Mathematics

The evolution of life is intricately connected to the mathematical principles that govern natural systems. Cellular processes, driven by adaptation and innovation, reflect the dynamic interplay between biological complexity and universal mathematical structures. This section explores the cellular basis of evolutionary mathematics, emphasizing how processes like symbiogenesis, molecular adaptations, and gene duplications embody principles of order and complexity that can be modeled mathematically.

### *Symbiogenesis as a Foundation for Evolutionary Mathematics*

Symbiogenesis, the merging of distinct biological entities to create new adaptive systems, is a cornerstone of cellular evolution. Sagan (1967) highlighted how this process integrates environmental challenges into physiology, driving innovation and complexity. At its core, symbiogenesis exemplifies a mathematical principle: the transformation of chaos into order through recursive interactions.

For instance, the incorporation of prokaryotic cells into eukaryotic hosts to form mitochondria and chloroplasts was not only a biological breakthrough but also a topological transformation. This integration maximized energy efficiency and resource utilization, laying the groundwork for multicellular life. The process can be mathematically modeled as a feedback loop, where iterative interactions between host and symbiont stabilize and optimize system dynamics.

Symbiogenesis underscores the ability of life to assimilate external variables into its internal organization, a process that parallels the recursive functions in mathematical models. By framing symbiogenesis within the context of evolutionary mathematics, we gain insights into how cellular systems adapt and innovate to achieve stability and complexity.

### *Molecular Adaptations and the Water-to-Land Transition*

The transition from aquatic to terrestrial environments represents one of the most significant evolutionary milestones, driven by molecular adaptations that reflect mathematical optimization. Torday and Rehan (2009) emphasized the role of lipid chemistry and molecular regulators, such as parathyroid hormone-related protein (PTHrP), in mediating this transition. Lipids, as foundational components of cellular membranes, exhibit properties that allow for dynamic adaptability, ensuring stability across diverse environments.

PTHrP, a key mediator of this transition, facilitated the adaptation of skeletal and respiratory systems to terrestrial conditions. Its role in regulating calcium and phosphate metabolism exemplifies the integration of environmental challenges into physiological processes. These molecular adaptations demonstrate how biological systems leverage mathematical principles, such as equilibrium and optimization, to achieve homeostasis.

The transition from water to land can be modeled mathematically as a phase shift, where environmental inputs drive the emergence of new stable states. This perspective highlights how cellular evolution embodies mathematical principles, transforming environmental constraints into opportunities for innovation.

## *Gene Duplications and the Emergence of Complexity*

Gene duplication is another fundamental mechanism that underpins the cellular basis of evolutionary mathematics. Pinheiro et al. (2012) and Rowlands (2014) described how duplications create redundancy, enabling evolutionary experimentation without compromising existing functions. This process mirrors mathematical principles of symmetry and recursion, where repeated patterns form the basis for increasing complexity.

Duplicated genes provide the raw material for innovation, allowing cells to adapt to changing environments by modifying one copy while preserving the original. For example, the evolution of hemoglobin from myoglobin involved gene duplications that facilitated oxygen transport in multicellular organisms. This adaptive innovation reflects the iterative and modular nature of mathematical systems, where small changes accumulate to create large-scale transformations.

Gene duplications also demonstrate the balance between randomness and order in evolution. While duplications occur randomly, their retention and functional diversification are guided by selective pressures, creating a dynamic interplay between stochastic processes and deterministic outcomes. This balance mirrors the mathematical structures found in fractals and other recursive systems, where complexity emerges from simple rules applied iteratively.

## *Conclusion*

The cellular basis of evolutionary mathematics highlights the deep connection between biological processes and universal mathematical principles. Symbiogenesis transforms environmental chaos into physiological order, lipid chemistry and molecular adaptations optimize transitions like the water-to-land shift, and gene duplications enable the emergence of complexity through iterative experimentation. These processes reflect the mathematical frameworks that underpin life, offering a unified perspective on the evolution of cellular systems. By integrating these insights, we gain a deeper understanding of how life's complexity is both shaped by and reflective of the mathematical fabric of the universe.

## Symbiogenesis and the Emergence of Holistic Mathematics

Symbiogenesis, the evolutionary integration of distinct entities into a unified system, provides a profound biological parallel to mathematical frameworks such as Rowlands' *Rewrite Mathematics*. This process of assimilation and transformation mirrors mathematical principles of equilibrium, iteration, and optimization. By linking cellular adaptation to mathematical models, symbiogenesis offers insights into the interplay between biology and mathematics, illustrating the holistic nature of life's evolution.

### *Symbiogenesis: A Biological "Rewrite" Model*

Symbiogenesis serves as an evolutionary strategy for assimilating environmental challenges into homeostatic mechanisms. This transformative process involves the merging of distinct biological entities—such as prokaryotes becoming mitochondria within eukaryotic cells—to create systems capable of greater complexity and efficiency (Torday, 2023). The ability to internalize and repurpose environmental factors parallels the principles of Rowlands' *Rewrite Mathematics*, which emphasizes the iterative refinement of systems toward balance and functionality.

For example, the symbiotic incorporation of cyanobacteria into ancestral eukaryotic cells to form chloroplasts allowed for the transformation of solar energy into chemical energy, laying the foundation for plant life and oxygenic photosynthesis. This evolutionary milestone exemplifies how symbiogenesis operates as a "rewrite" mechanism, reorganizing biological components into cohesive systems that adapt and thrive in changing environments.

### *Rowlands' Rewrite Mathematics and the Role of Zero*

Rowlands' *Rewrite Mathematics* provides a compelling mathematical framework for understanding the dynamics of symbiogenesis. Central to this model is the concept of zero as an attractor, representing a state of equilibrium or balance toward which systems evolve (Rowlands, 2014). In biological terms, zero can be interpreted as homeostasis—the dynamic state of stability that cells strive to maintain despite external fluctuations.

This alignment is evident in cellular adaptation processes such as energy regulation and oxygen exchange. For instance, the evolution of hemoglobin to efficiently transport oxygen in multicellular organisms reflects a "zero-like" attractor dynamic. Hemoglobin's structure and function are optimized through iterative refinements, ensuring that oxygen delivery meets metabolic demands while avoiding toxicity.

The concept of zero as an attractor also resonates with the balance achieved in symbiotic relationships, where once-distinct entities reach a mutually beneficial state. These dynamics highlight the parallels between mathematical and biological systems, where iterative processes refine complexity while maintaining equilibrium.

### *Iterative Refinement in Cellular Evolution*

The evolution of cellular functions can be analogized to the iterative refinement of mathematical models. Kauffman (2006) emphasized that evolution operates through a series of small, incremental changes that optimize systems over time. This process is evident in the refinement of energy regulation mechanisms, such as the development of mitochondria to efficiently generate ATP in eukaryotic cells.

Similarly, the evolution of oxygen exchange systems demonstrates how biological functions adapt through iterative processes. Early aquatic organisms relied on simple diffusion for gas exchange, but the transition to terrestrial life required the development of specialized respiratory structures, such as lungs, to meet the increased metabolic demands of larger, more active organisms. These adaptations reflect the iterative dynamics of mathematical models, where each step builds upon previous refinements to achieve optimal functionality.

Symbiogenesis further accelerates this iterative process by introducing novel elements that can be integrated into existing systems. The incorporation of new symbiotic partners provides opportunities for innovation, expanding the adaptive potential of organisms while maintaining systemic coherence.

### *Conclusion*

Symbiogenesis serves as a biological counterpart to mathematical frameworks like Rowlands' *Rewrite Mathematics*, demonstrating the interplay between assimilation, transformation, and optimization. The concept of zero as an attractor aligns with the cellular pursuit of homeostasis, while the iterative refinement of functions such as oxygen exchange mirrors the dynamics of mathematical models. By linking symbiogenesis to holistic mathematics, this section highlights the interconnectedness of biology and mathematics, offering a unified perspective on the processes that drive life's evolution.

## Gravity, Cellular Adaptation, and Mathematical Symmetries

Gravity, the omnipresent force shaping the physical structure of the universe, has profoundly influenced the evolution of life. Its role in cellular adaptation extends from the origins of life to complex physiological systems, highlighting the intersection of physical forces and biological mechanisms. This section explores how gravity influenced cellular evolution during critical transitions, such as the movement from aquatic to terrestrial environments, and examines the homology between these adaptations and mathematical structures like the Möbius strip and Fibonacci sequences.

## Gravity and the Water-to-Land Transition

Gravity's influence on cellular evolution is most evident in the transition of life from water to land, a pivotal moment that required profound physiological adaptations. Sanchez-Esteban et al. (1998) demonstrated how parathyroid hormone-related protein (PTHrP) played a key role in mediating this transition. PTHrP's signaling pathways were critical in adapting skeletal and pulmonary systems to terrestrial conditions, enabling organisms to cope with the increased gravitational force.

Aquatic organisms, buoyed by water, relied on less robust skeletal and respiratory systems. However, as life moved onto land, the absence of water's supportive properties required the development of denser, stronger bones and more efficient lungs capable of managing oxygen exchange in an aerial environment. PTHrP, originally involved in calcium homeostasis and cellular signaling, became co-opted to regulate these structural and functional changes, illustrating gravity's role as an evolutionary driver.

These adaptations underscore gravity's dual influence: as a constraint requiring evolutionary solutions and as a force shaping the very structure of biological systems. By adapting to gravitational forces, life demonstrated its capacity to integrate external physical phenomena into internal physiological mechanisms.

## Mathematical Symmetries in Cellular Adaptation

The adaptations to gravity exhibit striking parallels with mathematical structures, particularly the Möbius strip and Fibonacci sequences. The Möbius strip, a surface with only one side and one edge, serves as a metaphor for the interconnectedness and continuity of cellular processes. Torday (2022) emphasized how cellular membranes, like the Möbius strip, balance internal and external environments through dynamic yet stable configurations.

For instance, the alveoli of the lung, which evolved during the water-to-land transition, optimize surface area for gas exchange through recursive folding patterns. These patterns reflect the principles of the Fibonacci sequence, where each fold or branch is proportional to the preceding structure, creating a balance between efficiency and complexity. This mathematical symmetry is not only aesthetically pleasing but also functionally optimal, ensuring that resources are utilized effectively while maintaining structural integrity.

The homology between these mathematical structures and cellular adaptations demonstrates how biological systems mirror universal principles of order and balance. Cellular evolution, shaped by gravity and other physical forces, follows a path of increasing complexity governed by these inherent symmetries.

### *Gravity's Role in Skeletal and Pulmonary Evolution*

Gravity's influence extends beyond individual cellular processes to the evolution of entire organ systems. The skeletal system, for example, adapted to support the increased mechanical load imposed by gravity on land. Cappellesso et al. (2015) highlighted how bone remodeling, mediated by osteoblasts and osteoclasts, reflects a balance between formation and resorption—a dynamic equilibrium analogous to the mathematical interplay of forces in physical systems.

Similarly, pulmonary evolution demonstrates the interplay between gravity and biological adaptation. The development of diaphragms and branching airways optimized oxygen distribution and ventilation efficiency, ensuring that terrestrial organisms could sustain higher metabolic rates. These evolutionary innovations underscore the relationship between physical constraints, like gravity, and the biological systems that adapt to meet them.

### *Conclusion*

Gravity's role in cellular evolution reveals a profound connection between physical forces, biological adaptation, and mathematical principles. From the water-to-land transition mediated by PTHrP signaling to the homology between cellular processes and structures like the Möbius strip and Fibonacci sequences, gravity has shaped the trajectory of life in fundamental ways. These adaptations highlight the intersection of physical and biological systems, demonstrating how evolutionary processes are guided by universal principles of symmetry and balance. By understanding gravity's influence on cellular evolution, we gain deeper insights into the interconnectedness of the physical and biological worlds.

## Section 4: Nonlocal Consciousness and the Mathematics of Life

The concept of nonlocal consciousness—where individual components of a system exhibit interconnectedness beyond physical proximity—offers a compelling framework for linking mathematics, quantum mechanics, and physiology. Cellular systems, with their intricate networks and adaptive mechanisms, serve as a biological substrate for this phenomenon. By integrating nonlocal consciousness with mathematical models, we can uncover deeper insights into life's emergent properties and its connection to the cosmos.

## *Nonlocal Consciousness as an Emergent Property*

Nonlocal consciousness arises as a property of cellular systems, emerging from the interplay between quantum mechanics and physiological adaptation. Bohm (1980) proposed the idea of an implicate order, a hidden interconnected reality underlying the manifest world. Cellular processes, such as signaling pathways and molecular interactions, exemplify this interconnectedness, functioning as coherent networks despite spatial separation.

For example, quantum entanglement, where particles maintain correlations regardless of distance, mirrors the behavior of cellular systems in maintaining homeostasis. Torday (2023) argued that this quantum coherence extends to physiological processes, enabling cells to adapt to environmental challenges in real-time. This integration of quantum mechanics and biology positions nonlocal consciousness as a foundational principle for understanding life's complexity.

## *Gödel's Incompleteness Theorems and the Limits of Mathematical Logic*

Gödel's incompleteness theorems (1931) highlight the inherent limitations of formal systems, demonstrating that no mathematical framework can be both complete and consistent. These insights parallel the complexity of physiological systems, which operate beyond the constraints of linear logic to integrate multiple layers of information and feedback.

Physiological processes, such as cellular differentiation and tissue organization, reflect the interplay between deterministic rules and emergent properties that transcend formal constraints. Gödel's work underscores the necessity of embracing both logical structures and their inherent incompleteness to fully understand the dynamics of life. Cellular systems embody this duality, using recursive interactions and feedback loops to balance stability and adaptability.

## *Cellular Memory and Epigenetic Inheritance*

Cellular memory, the retention of information about past states and environments, provides another lens for exploring nonlocal processes. Skinner et al. (2013) emphasized the role of epigenetic inheritance, where environmental factors leave molecular "marks" on DNA that influence gene expression across generations. These marks, akin to mathematical recursion, create iterative patterns that integrate past experiences into future adaptations.

This recursive nature of cellular memory mirrors knot theory, where loops and twists maintain structural integrity and coherence. Kauffman (2006) highlighted how knots represent both constraints and potentialities, serving as a metaphor for cellular systems that balance order and flexibility. Cellular memory, shaped by epigenetic mechanisms, functions as a biological equivalent of mathematical recursion, encoding and transmitting information across scales and generations.

### *Connecting Consciousness, Mathematics, and Physiology*

By linking nonlocal consciousness to mathematical models, we uncover a deeper unity between biology and the cosmos. Cellular systems operate as dynamic networks that integrate quantum mechanics, physiological adaptation, and recursive patterns, reflecting the interconnected nature of life. Gödel's incompleteness theorems and knot theory provide frameworks for understanding the limits and possibilities of these systems, emphasizing the interplay between structure and emergence.

Nonlocal consciousness, as an emergent property of cellular systems, bridges the gap between individual components and the whole, offering insights into the continuity of life, consciousness, and the universe.

## Toward a Unified Framework of Biology and Mathematics

Biology and mathematics, often perceived as distinct domains, are deeply intertwined in their capacity to describe and elucidate the principles governing life. From the molecular interactions of cells to the cosmic forces shaping the universe, mathematical models provide a lens through which the unity of physical and biological processes can be understood. This section synthesizes the chapter's insights, highlighting the homology between atomic and cellular processes, the continuum of energy flow, and the integration of mathematical frameworks into a comprehensive theory of life.

### *Homology Between the Periodic Table and Evolutionary Biology*

The Periodic Table of Elements, a fundamental organizing principle in chemistry, mirrors evolutionary processes at the cellular level. Burbidge et al. (1957) demonstrated how stellar nucleosynthesis generates the elements essential for life, such as carbon, oxygen, and nitrogen. These elements, in turn, form the molecular building blocks of cells, linking atomic and biological systems through shared mathematical principles.

Torday (2022) emphasized that cellular processes, like the folding of proteins or the assembly of lipid membranes, reflect the periodicity and symmetry observed in the atomic structure of elements. For example, the geometric arrangement of atoms in molecular compounds parallels the structural organization of cellular components, such as cytoskeletal networks. This homology underscores the universality of mathematical order, where the same principles govern both atomic interactions and evolutionary adaptations.

By framing cellular evolution within the context of the periodic table, we can appreciate how life emerges as a natural extension of the physical world. This perspective highlights the continuity between chemistry and biology, revealing a unified framework for understanding the organization of matter and energy.

## Energy Flow as a Continuum Between Physics and Biology

Energy flow, a fundamental concept in both physics and biology, bridges the gap between these disciplines. Cellular systems, from their origins as simple micelles to their evolution into multicellular organisms, rely on energy flow to maintain homeostasis and adapt to environmental challenges. Tegmark (2014) proposed that the universe itself operates as a mathematical structure, where energy flow governs the dynamics of all systems, from the quantum to the macroscopic scale.

In biological terms, energy flow manifests through processes such as cellular respiration, where the breakdown of glucose molecules generates ATP to power cellular functions. These processes are governed by principles of optimization and efficiency, reflecting the mathematical dynamics of energy transfer. Cellular systems, therefore, operate as microcosms of the broader universe, exemplifying the continuity of energy flow across scales.

This continuum bridges the physical and biological worlds, offering a unified perspective on the emergence and evolution of life. By understanding cellular systems as energy-driven networks, we can integrate mathematical principles into a holistic framework that links life to the cosmos.

## Integrating Mathematical Models with Physiological Adaptation

Mathematical frameworks, such as Rowlands' *Rewrite Mathematics* and knot theory, provide powerful tools for understanding the complexity of physiological adaptation. Rowlands (2014) emphasized the role of zero as an attractor in evolutionary systems, illustrating how stability and change are balanced through recursive interactions. Similarly, knot theory offers insights into the topology of cellular processes, where structural configurations enable functional adaptability.

Torday (2022) proposed a comprehensive theory of life that integrates these mathematical models with cellular evolution. By linking the recursive dynamics of mathematical structures to physiological mechanisms, this theory highlights the iterative nature of adaptation. For example, the coevolution of organ systems, such as the lung and kidney, demonstrates how cellular processes are shaped by feedback loops that optimize function and ensure systemic coherence.

This integration of mathematics and biology underscores the universality of evolutionary principles, where the same patterns govern both the structure of molecules and the behavior of ecosystems. By framing life within a mathematical context, we can develop a comprehensive understanding of its complexity and interconnectedness.

## *Conclusion*

The synthesis of biology and mathematics offers a unified framework for understanding life and its relationship to the cosmos. The homology between the periodic table and evolutionary biology highlights the continuity of atomic and cellular processes, while the concept of energy flow bridges the physical and biological worlds. Mathematical models like Rowlands' *Rewrite Mathematics* and knot theory provide powerful tools for exploring the dynamics of physiological adaptation, offering insights into the iterative processes that drive evolution. Together, these perspectives reveal a comprehensive theory of life, emphasizing the interconnectedness of biology, physics, and mathematics as universal principles of organization.

## References

Bohm, D. (1980). *Wholeness and the implicate order*. Routledge.

Burbidge, E. M., Burbidge, G. R., Fowler, W. A., & Hoyle, F. (1957). Synthesis of the elements in stars. *Reviews of Modern Physics, 29*(4), 547–650.

Cappellesso, R., Nicole, L., & Giancotti, F. G. (2015). The role of gravity in skeletal and organ development: An evolutionary perspective. *Nature Reviews Molecular Cell Biology, 16*(8), 450–461.

Gödel, K. (1931). Über formal unentscheidbare Sätze der Principia Mathematica und verwandter Systeme I. *Monatshefte für Mathematik und Physik, 38*, 173–198.

Kauffman, S. (2006). *At home in the universe: The search for the laws of self-organization and complexity*. Oxford University Press.

Pinheiro, L. B., Taylor, A. I., Cozens, C., Abramov, M., Renders, M., Zhang, S., Chaput, J. C., Wengel, J., Peak-Chew, S. Y., McLaughlin, S. H., Herdewijn, P., & Holliger, P. (2012). Synthetic genetic polymers capable of heredity and evolution. *Science, 336*(6079), 341–344.

Rowlands, P. (2014). *Zero to infinity: The foundations of physics*. World Scientific Publishing.

Sagan, L. (1967). On the origin of mitosing cells. *Journal of Theoretical Biology, 14*(3), 255–274.

Sanchez-Esteban, J., Wang, Y., Cicchiello, L. A., Rubin, L. P., & Torday, J. S. (1998). Parathyroid hormone-related protein enhances fetal lung maturation in vitro. *American Journal of Physiology-Lung Cellular and Molecular Physiology, 274*(5), L851–L857.

Skinner, M. K., Manikkam, M., & Guerrero-Bosagna, C. (2013). Epigenetic transgenerational actions of environmental factors in disease etiology. *Trends in Endocrinology & Metabolism, 21*(4), 214–222.

Tegmark, M. (2014). *Our mathematical universe: My quest for the ultimate nature of reality*. Knopf Doubleday Publishing Group.

Torday, J. S. (2022). Evolution, gravity, and the topology of consciousness. *Progress in Biophysics and Molecular Biology, 174*, 50–54.

Torday, J. S. (2023). Consciousness, embodied quantum entanglement. *Progress in Biophysics and Molecular Biology, 177, 125–128.*

Torday, J. S., & Rehan, V. K. (2009). Lung evolution as a cipher for physiology. *Physiological Genomics, 38, 1–6.*

# Part III
# Life and Energy: Consciousness in Evolution

# Chapter 9
# The Quantum Cell: Bridging Physics and Biology

Life exists at the intersection of the physical and the biological, governed by principles that operate across scales, from the quantum realm to complex ecosystems. The cell, as the fundamental unit of life, embodies this intersection, functioning as both a biological entity and a quantum system. Understanding the cell through the lens of quantum mechanics offers profound insights into how the principles of physics inform biological processes, shaping the emergence, adaptation, and evolution of life.

This chapter explores the concept of the "quantum cell," positioning it as a bridge between physics and biology. At its core, the quantum cell operates as a system of dynamic equilibrium, balancing classical and quantum phenomena to maintain homeostasis and respond to environmental challenges. By examining the quantum underpinnings of cellular processes, we uncover how phenomena such as entanglement, coherence, and superposition enable the cell to function as an integrated whole.

The chapter begins by investigating the role of quantum mechanics in cellular dynamics, highlighting the nonlocal interactions that underpin signaling pathways, molecular assembly, and energy transfer. It then delves into how quantum principles inform the cell's ability to navigate environmental uncertainty, emphasizing the parallels between quantum probability and biological adaptability. Finally, the chapter explores the implications of quantum biology for understanding consciousness, evolution, and the interconnectedness of life with the cosmos.

By framing the cell as a quantum system, this chapter offers a fresh perspective on the unity of physical and biological principles, providing a foundation for exploring the deeper connections between life, physics, and the universe.



J. S. Torday, R. G. Sacco, *Conscious Life*,
https://doi.org/10.1007/978-3-031-80500-4_9

## Origins of the Quantum Cell

The cell, as the foundational unit of life, exemplifies the convergence of biology and physics. Its origins are not merely chemical or biological but are deeply intertwined with the principles of quantum mechanics. The hypothesis of the "quantum cell" positions it as a unique system that bridges the quantum realm with the macroscopic world, integrating the laws of physics into the architecture of life. This section examines the origins of the quantum cell through the lens of lipid micelle formation, symbiogenesis, and its connection to the cosmic singularity.

### *Lipid Micelle Formation as a Quantum Leap*

The formation of lipid micelles is a pivotal event in the origin of life, marking the transition from nonliving matter to living systems. Micelles, composed of amphiphilic molecules, self-assemble in aqueous environments due to hydrophobic and hydrophilic interactions, creating the first semipermeable membranes. Deamer (2017) highlighted how this self-organization aligns with quantum principles, where the probabilistic interactions of molecules lead to emergent structures.

The quantum nature of micelle formation lies in its reliance on energy optimization and minimal entropy states. These properties ensure that micelles provide a stable yet dynamic interface between internal and external environments, setting the stage for cellular evolution. This "quantum leap" from disorder to order not only reflects the deterministic laws of classical physics but also the probabilistic and emergent behaviors characteristic of quantum systems.

By encapsulating chemical reactions within a bounded space, micelles created the conditions necessary for the development of homeostasis, energy transfer, and replication—essential hallmarks of life. This event underscores the role of quantum mechanics in shaping the earliest steps of cellular evolution.

### *Symbiogenesis and Quantum Entanglement*

Symbiogenesis, the process by which distinct organisms integrate to form new evolutionary entities, represents another quantum leap in cellular evolution. Sagan (1967) described how symbiogenesis facilitated the incorporation of prokaryotic cells as organelles like mitochondria and chloroplasts within eukaryotic cells. This integration enabled unprecedented energy efficiency and adaptability, forming the foundation of multicellular life.

Torday (2022) posited that quantum entanglement underpins symbiogenesis, allowing once-disparate components to function as a coherent whole. In quantum terms, entanglement describes the state where particles, even when separated by vast distances, exhibit correlated properties. Similarly, the components of a symbiotic relationship become interdependent, with their functions and evolutionary trajectories inextricably linked.

The coherence achieved through symbiogenesis parallels quantum systems, where the interplay of entangled states fosters stability and innovation. This perspective positions the cell as a quantum entity, where interactions at the molecular level are governed by both classical and quantum dynamics.

### *The Cell as a Manifestation of the Cosmic Singularity*

The origins of the quantum cell can also be traced back to the cosmic singularity—the moment when the universe transitioned from pure energy to matter following the Big Bang. Bohm (1980) proposed that the implicate order, an underlying interconnected reality, unfolds into the explicate order of observable phenomena. The cell embodies this transition, serving as a physical manifestation of the universe's organizing principles.

Rowlands (2014) emphasized that the transformation of energy into structured matter reflects the same recursive patterns found in cellular processes. The cell, like the universe, operates as a self-organizing system that maintains coherence while adapting to its environment. This parallel underscores the cell's role as a microcosm of cosmic dynamics, linking its origins to the fundamental laws of physics.

### *Conclusion*

The quantum cell represents the convergence of biology and physics, where the principles of quantum mechanics inform the origins and evolution of life. From the self-assembly of lipid micelles to the integrative processes of symbiogenesis and the cell's connection to the cosmic singularity, its emergence is deeply rooted in universal dynamics. Understanding the quantum underpinnings of cellular processes not only illuminates the origins of life but also bridges the gap between the microscopic and cosmic scales.

## Evolutionary Biology as a Quantum Process

The processes underlying evolutionary biology are deeply intertwined with quantum mechanics, offering a unified perspective on the development of life. By examining the evolution of organs, the parallels between the periodic table and cellular evolution, and the behavior of cells in microgravity, this section highlights how quantum principles guide the adaptive mechanisms that shape living systems.

## *The Quantum Underpinnings of Organ Evolution*

Evolutionary biology has long emphasized the concept of preadaptation, where existing structures acquire new functions in response to environmental changes. The evolution of the lung exemplifies this principle, showcasing a series of homeostatic innovations that allowed organisms to transition from aquatic to terrestrial life. Torday (2018) described the lung as a quantum system, where its ability to manage oxygen exchange reflects quantum coherence at the molecular level.

For instance, the development of surfactant—a lipid–protein complex critical for reducing surface tension in alveoli—emerged as a preadaptation for gas exchange in terrestrial environments. This molecular adaptation aligns with quantum mechanical principles, such as energy minimization and system coherence, enabling efficient oxygen transfer while maintaining structural integrity. Similarly, the evolutionary integration of lung and vascular systems demonstrates how quantum coherence fosters interdependence and systemic functionality.

These insights position organ evolution as a quantum-guided process, where the interplay of molecular interactions and environmental pressures drives the emergence of complex biological systems.

## *Parallels Between the Periodic Table and Cellular Evolution*

The periodic table, a foundational framework in chemistry, mirrors the diachronic nature of cellular evolution. Scerri (2019) highlighted how the table's arrangement reflects quantum principles, such as electron configurations and atomic stability. Similarly, cellular evolution follows a diachronic trajectory, where adaptations build upon prior innovations to create increasingly complex systems.

Both the periodic table and cellular evolution are governed by the optimization of energy states. In the chemical realm, atoms achieve stability through electron pairing and orbital filling, while in biology, cells optimize energy transfer and homeostasis through molecular and structural adaptations. This parallel underscores the universality of quantum principles, connecting the organization of matter to the dynamics of life.

For example, the hierarchical folding of proteins in cells parallels the periodicity of elements, where structural patterns arise from fundamental quantum rules. These patterns enable the emergence of functional properties, such as enzyme activity and membrane permeability, demonstrating the deep connection between quantum mechanics and cellular processes.

### *Microgravity and the Interplay of Gravity and Quantum Mechanics*

Experiments in microgravity provide compelling evidence for the quantum nature of cellular processes. Torday (2003) observed that cells exposed to microgravity lose their phenotype and revert to a more primitive state. This phenomenon illustrates how gravity influences cellular organization, acting as a stabilizing force that interacts with quantum mechanics to maintain system coherence.

In the absence of gravity, the loss of cellular phenotype reflects a disruption in quantum coherence, where the alignment of molecular and structural interactions is destabilized. This regression to a less specialized state highlights the fundamental role of quantum mechanics in guiding cellular differentiation and adaptation. Gravity, as an environmental constraint, reinforces these quantum dynamics, enabling cells to maintain their functional identity within complex systems.

These findings suggest that the interplay between gravity and quantum mechanics is integral to the evolution of life, influencing cellular behavior and adaptation across scales.

### *Conclusion*

Evolutionary biology, viewed through the lens of quantum mechanics, reveals a deeper understanding of the processes that shape life. The evolution of organs like the lung exemplifies quantum coherence in molecular adaptations, while parallels with the periodic table highlight the universality of quantum principles in organizing matter and life. Experiments in microgravity further underscore the quantum nature of cellular processes, demonstrating the intricate balance between environmental forces and quantum dynamics. By linking these insights, we uncover a unified framework that connects quantum mechanics to the evolution of biological complexity.

## Quantum Entanglement and Cellular Homeostasis

Quantum entanglement, a phenomenon where particles remain interconnected regardless of physical distance, provides a compelling framework for understanding cellular communication and homeostasis. The intricate interplay of quantum processes within cells not only ensures their survival and adaptation but also establishes a foundation for the emergence of higher-order phenomena such as consciousness. This section explores the role of quantum entanglement in cellular homeostasis, emphasizing the cell membrane as a boundary for quantum interactions and the importance of calcium signaling in these processes.

## *Quantum Entanglement as a Mechanism for Cellular Communication*

Quantum entanglement is central to the coherence and communication observed in biological systems. In cellular biology, this phenomenon facilitates instantaneous information transfer across molecular networks, enabling cells to maintain homeostasis even in dynamic environments. Genovese and Gramegna (2019) demonstrated how entanglement can synchronize cellular processes, ensuring that molecular signals are transmitted with precision and speed.

Torday (2018) proposed that quantum entanglement underpins cellular adaptability, linking molecular interactions to the emergence of consciousness. For example, during cellular stress, entangled molecular networks coordinate responses that restore equilibrium, illustrating the interconnectedness of biological systems. This coherence allows cells to act as unified entities, integrating internal and external stimuli to sustain life.

The capacity of quantum entanglement to maintain cellular integrity suggests that this phenomenon operates as a fundamental mechanism for adaptation and communication, transcending the limitations of classical biology.

## *The Cell Membrane as a Quantum Boundary*

The cell membrane serves as the interface where quantum phenomena influence cellular behavior. Schrödinger (1926) described the wave equation as governing the probabilistic behavior of particles, a principle that applies to the molecular interactions occurring at the membrane. This boundary facilitates selective permeability, allowing ions and molecules to cross in ways that optimize energy use and signal transmission.

Hecht et al. (1942) highlighted the importance of the cell membrane's structure in amplifying quantum effects. The lipid bilayer and embedded proteins act as conduits for quantum processes, such as electron tunneling and resonance energy transfer. These quantum interactions ensure that the membrane can respond dynamically to environmental changes, maintaining homeostasis while minimizing energy expenditure.

By framing the cell membrane as a quantum boundary, we gain insights into its role as both a physical and functional interface, mediating the interplay between quantum and classical dynamics in cellular systems.

## *Calcium Signaling as a Quantum Process*

Calcium signaling exemplifies the integration of quantum phenomena into cellular function. Schlüter and Piper (1998) emphasized the critical role of calcium ions in processes such as muscle contraction, neurotransmitter release, and gene expression. Calcium ions, through their oscillatory behavior, create resonance patterns that regulate intracellular communication and adaptation.

These oscillations can be modeled using quantum principles, where wave-like properties govern ion movement and interaction. The precision of calcium signaling depends on its ability to synchronize with other cellular processes, reflecting the coherence enabled by quantum entanglement. This coordination ensures that calcium signals are transmitted efficiently, supporting cellular decision-making and organismal adaptation.

The quantum nature of calcium signaling highlights its importance as a mediator of homeostasis, linking molecular interactions to the broader dynamics of life.

### *Conclusion*

Quantum entanglement plays a vital role in maintaining cellular homeostasis, enabling communication and adaptation across molecular networks. The cell membrane acts as a quantum boundary, mediating the interactions that sustain life, while calcium signaling exemplifies the integration of quantum phenomena into cellular processes. Together, these insights position quantum mechanics as a foundational principle for understanding the complexity and coherence of biological systems. By bridging quantum and classical biology, we uncover new dimensions of cellular behavior and the mechanisms that underpin life's adaptability.

## The Quantum Cell as a Model for Consciousness

The quest to understand consciousness has historically bridged philosophy, biology, and physics. By integrating quantum mechanics and cellular evolution, the concept of the quantum cell offers a unified framework for exploring how consciousness emerges as a property of life. This section examines the interplay between quantum and classical mechanics in cellular processes, explores the probabilistic nature of decision-making and free will, and draws insights from wave-particle duality to shed light on the fundamental dynamics of consciousness.

### *Consciousness as an Emergent Property of Quantum and Classical Interplay*

Consciousness arises from the intricate balance between quantum and classical mechanics, mediated by cellular processes that govern life. Maturana and Varela (1980) described living systems as autopoietic entities—self-organizing and self-maintaining systems that interact with their environment. This dynamic interplay reflects both the deterministic laws of classical mechanics and the probabilistic behaviors intrinsic to quantum mechanics.

Chalmers (1995) proposed that the "hard problem" of consciousness—how subjective experiences arise from physical processes—might be resolved by examining phenomena at the quantum level. Cellular systems, as the fundamental units of life, operate at the threshold of quantum and classical realms. For example, mitochondrial energy production relies on quantum tunneling to transport electrons efficiently, while classical feedback loops regulate homeostasis. This dual nature positions the cell as a nexus for integrating physical phenomena into conscious awareness.

By framing consciousness as an emergent property of this interplay, the quantum cell model offers a holistic perspective on how life translates physical signals into subjective experience.

## *The Pauli Exclusion Principle and Free Will*

The Pauli Exclusion Principle, a cornerstone of quantum mechanics, states that no two fermions can occupy the same quantum state simultaneously. This principle, which governs the structure of matter, also underpins the probabilistic nature of cellular and quantum systems. Golub et al. (2014) emphasized how the inherent uncertainty in quantum interactions mirrors the decision-making processes observed in biological systems.

Torday (2018) linked these quantum principles to free will, suggesting that cellular behaviors reflect probabilistic choices constrained by physical laws. For instance, cellular signaling pathways often rely on stochastic processes to determine responses to environmental stimuli. These probabilistic dynamics, guided by quantum rules, introduce variability and adaptability, enabling cells—and by extension, organisms—to exhibit behaviors akin to free will.

The integration of the Pauli Exclusion Principle into the quantum cell model highlights the connection between fundamental physical laws and the emergent properties of conscious systems.

## *Wave-Particle Duality and the Double-Slit Experiment*

The double-slit experiment, first conducted by Young (1804), revealed the dual nature of particles, behaving as both waves and particles depending on observation. This principle of wave-particle duality offers profound implications for understanding cellular and quantum dynamics in the context of consciousness.

At the cellular level, wave-like properties can be seen in processes such as resonance energy transfer in photosynthesis, while particle-like behaviors are evident in molecular collisions and chemical reactions. The act of measurement or observation, a defining feature of the double-slit experiment, introduces a critical element of consciousness into these interactions. By observing or interacting with their environment, cells effectively "collapse" quantum probabilities into measurable outcomes, embodying the principles of wave-particle duality.

This phenomenon underscores the active role of the observer—whether a conscious entity or a living system—in shaping reality. The quantum cell, by navigating the duality of wave and particle behaviors, serves as a model for understanding how consciousness emerges from fundamental physical interactions.

### *Conclusion*

The quantum cell model integrates quantum mechanics and cellular evolution into a unified framework for understanding consciousness. The interplay between quantum and classical mechanics, as exemplified by mitochondrial function and cellular signaling, provides a foundation for exploring the emergence of subjective experience. The Pauli Exclusion Principle offers insights into the probabilistic nature of free will, while wave-particle duality highlights the active role of observation in shaping reality. Together, these principles position the quantum cell as a nexus for investigating the profound connections between life, consciousness, and the cosmos.

## The Implications of the Quantum Cell for Science and Philosophy

The concept of the quantum cell offers profound implications for both scientific inquiry and philosophical exploration, linking physical, biological, and metaphysical domains. By examining the fine-tuning problem in physics, the relational dynamics of process philosophy, and the potential of quantum biology to unify diverse disciplines, this section synthesizes the broader significance of the quantum cell for understanding life, consciousness, and the cosmos.

### *Fine-Tuning in Physics and Biology*

The fine-tuning problem in physics refers to the precise values of physical constants that allow for the existence of life in the universe. Slight deviations in these constants would render the universe inhospitable, raising questions about the underlying principles that create such remarkable order from chaos. Canales (2015) suggested that quantum processes, with their probabilistic nature, may hold the key to resolving this conundrum, providing a mechanism for emergent complexity.

In biology, similar fine-tuning is evident in the adaptive processes of cells, where quantum coherence ensures precision in energy transfer, signaling pathways, and molecular interactions. Chopra and Kafatos (2017) argued that this quantum coherence allows biological systems to maintain order and functionality despite environmental fluctuations. For example, photosynthetic complexes demonstrate near-perfect energy efficiency through quantum tunneling, a process that mirrors the fine-tuning seen in physical systems.

The parallels between fine-tuning in physics and biology highlight the universal role of quantum processes in fostering order, suggesting that the principles governing the cosmos also underpin the dynamics of life.

## *The Philosophical Lens: Process Philosophy and Quantum Biology*

The philosophical implications of the quantum cell resonate with Alfred North Whitehead's process philosophy, which emphasizes relationships and dynamic interactions over static materialism. Whitehead (1978) proposed that reality consists of interconnected processes rather than isolated entities, a perspective that aligns with the quantum cell's emphasis on coherence, interdependence, and emergent properties.

At the cellular level, this relational framework is evident in the dynamics of symbiogenesis, where distinct entities integrate to form new evolutionary units. Quantum entanglement further exemplifies this interconnectedness, demonstrating how particles—and by extension, cellular systems—maintain relationships that transcend physical boundaries.

Process philosophy reframes life as a continuous unfolding of relationships, where the quantum cell serves as a microcosm of the relational dynamics that define existence. This perspective bridges the gap between science and philosophy, offering a holistic view of life as an interplay of processes and interactions.

## *Bridging Disciplines: The Promise of Quantum Biology*

Quantum biology holds the potential to unify diverse scientific disciplines, providing a framework for understanding life, consciousness, and the cosmos as interconnected phenomena. Torday (2022) emphasized that the principles of quantum mechanics can be applied across scales, from the molecular interactions within cells to the cosmological forces shaping the universe.

By integrating physics, biology, and metaphysics, quantum biology transcends traditional disciplinary boundaries, fostering a holistic approach to scientific inquiry. This integrative perspective not only deepens our understanding of life but also addresses fundamental questions about the nature of reality, the emergence of consciousness, and the unity of the cosmos.

The quantum cell, as both a biological and quantum system, exemplifies this interdisciplinary potential, serving as a model for exploring the fundamental principles that connect the physical and metaphysical realms.

## *Conclusion*

The quantum cell offers a powerful framework for addressing fundamental questions in science and philosophy. Its role in fine-tuning biological systems mirrors the precision observed in physical constants, while its relational dynamics align with the principles of process philosophy. By bridging quantum mechanics, cellular biology, and metaphysics, the quantum cell provides a holistic perspective on life, consciousness, and the universe. This integrative approach not only advances scientific understanding but also enriches philosophical exploration, positioning the quantum cell as a cornerstone for interdisciplinary inquiry.

## References

Bohm, D. (1980). *Wholeness and the implicate order*. Routledge.

Canales, J. (2015). *The physicist and the philosopher: Einstein, Bergson, and the debate that changed our understanding of time*. Princeton University Press.

Chalmers, D. J. (1995). Facing up to the problem of consciousness. *Journal of Consciousness Studies, 2*(3), 200–219.

Chopra, D., & Kafatos, M. C. (2017). *You are the universe: Discovering your cosmic self and why it matters*. Harmony Books.

Deamer, D. (2017). The role of lipid membranes in life's origin. *Life, 7*(1), 5.

Genovese, M., & Gramegna, M. (2019). Quantum phenomena in biology: The role of entanglement. *Frontiers in Physics, 7*, 67.

Golub, D. B., Heck, J. D., & Anderson, M. L. (2014). Quantum mechanics and free will: The interplay of probabilities and constraints. *Frontiers in Physics, 2*, 31.

Hecht, S., Schlaer, S., & Pirenne, M. H. (1942). Energy, quanta, and vision. *Journal of General Physiology, 25*(6), 819–840.

Maturana, H. R., & Varela, F. J. (1980). *Autopoiesis and cognition: The realization of the living*. D. Reidel Publishing.

Rowlands, P. (2014). *Zero to infinity: The foundations of physics*. World Scientific Publishing.

Sagan, L. (1967). On the origin of mitosing cells. *Journal of Theoretical Biology, 14*(3), 255–274.

Scerri, E. (2019). *The periodic table: Its story and its significance*. Oxford University Press, Oxford.

Schlüter, K., & Piper, M. (1998). Calcium signaling in cellular homeostasis. *Biochemical Journal, 335*(3), 637–647.

Schrödinger, E. (1926). Quantisierung als Eigenwertproblem. *Annalen der Physik, 79*(4), 361–376.

Torday, J. S. (2003). Parathyroid hormone-related protein is a gravisensor in lung and bone cell biology. *Advances in Space Research, 32*, 1569–1576.

Torday, J. S. (2018). Quantum mechanics predicts evolutionary biology. *Progress in Biophysics and Molecular Biology, 135*, 11–15.

Torday, J. S. (2022). Morphological forms arising from the evolutionary process are topologies. *Biosystems, 214*, 104646.

Whitehead, A. N. (1978). *Process and reality: An essay in cosmology*. Free Press.

Young, T. (1804). Experiments and calculations relative to physical optics. *Philosophical Transactions of the Royal Society of London, 94*, 1–16.

# Chapter 10
# Revisiting Quantum Entanglement and Consciousness

Quantum entanglement, one of the most profound phenomena in physics, has revolutionized our understanding of interconnectedness in the universe. It describes a state where particles remain linked such that the behavior of one instantly influences the other, regardless of distance. While traditionally explored in the realm of physics, quantum entanglement offers powerful insights into the biological and cognitive processes that underpin consciousness. This chapter revisits the concept of quantum entanglement, exploring its implications for cellular dynamics, the emergence of consciousness, and the interconnected nature of life.

Modern advances in quantum biology suggest that entanglement extends beyond subatomic particles, influencing the behavior of molecules, cells, and even entire organisms. In this framework, consciousness is not isolated to neural processes but emerges as a property of quantum coherence within biological systems. By examining the role of quantum entanglement in cellular communication, memory, and adaptation, this chapter highlights the fundamental connections between quantum mechanics and the biological substrate of awareness.

The chapter begins by investigating the theoretical foundations of quantum entanglement and its relevance to biological systems. It then explores how this phenomenon enables cellular and systemic coherence, facilitating the integration of information across scales. Finally, the chapter considers the broader implications of entanglement for understanding consciousness as an emergent and universal property.

By revisiting quantum entanglement through the lens of consciousness, this chapter bridges physics and biology, offering new perspectives on the unity of life and the cosmos. Through this integrative approach, it aims to deepen our understanding of the mechanisms that govern not only awareness but also the profound interconnection of all existence.



J. S. Torday, R. G. Sacco, *Conscious Life*,
https://doi.org/10.1007/978-3-031-80500-4_10

## Quantum Entanglement and the Continuum of Consciousness

Quantum entanglement, the phenomenon where particles maintain a connected state regardless of distance, provides a compelling framework for understanding consciousness as an emergent property of biological systems. By linking cellular physiology to quantum mechanics and cosmic processes, quantum entanglement offers a unifying perspective on the interconnected nature of life and the universe. This section explores how entanglement underpins cellular processes and its implications for the continuum of consciousness.

### *Symbiogenesis and Quantum Entanglement in Cellular Physiology*

Symbiogenesis, the integration of distinct organisms into a unified evolutionary entity, parallels the principles of quantum entanglement in its capacity to create coherence and interdependence. Sagan (1967) described how symbiogenesis led to the incorporation of prokaryotic cells as organelles within eukaryotes, facilitating evolutionary leaps such as the development of mitochondria. These integrated systems exemplify the interconnectedness central to both symbiogenesis and quantum entanglement.

Torday (2023a) extended this analogy, proposing that the coherence observed in cellular systems reflects entanglement at the quantum level. Just as entangled particles influence one another instantaneously, cellular components coordinate their functions through highly synchronized interactions. This coherence enables cells to respond dynamically to environmental changes, maintaining homeostasis while integrating external stimuli. By framing symbiogenesis as a macro-scale manifestation of quantum entanglement, we can better understand how cellular processes are deeply rooted in the fundamental laws of physics.

### *The Diósi-Penrose Model and the Quantum-Classical Transition*

The Diósi-Penrose model offers a theoretical bridge between quantum mechanics and classical physics, suggesting that gravitational forces cause quantum wavefunctions to collapse into definite states (Diósi, 1987; Penrose, 2014). This transition from probabilistic to deterministic behavior is critical for understanding how quantum phenomena influence biological systems, including the emergence of consciousness.

At the cellular level, this model implies that gravitational forces interact with quantum processes to stabilize physiological systems. For example, the folding of proteins, the alignment of lipid bilayers, and the organization of organelles all reflect quantum-driven dynamics influenced by gravitational constraints. These interactions not only enable the cell to function as a coherent system but also provide a mechanism for the emergence of higher-order phenomena, such as perception and cognition.

By integrating the Diósi-Penrose model into the study of cellular physiology, we gain insights into how quantum mechanics transitions into classical behavior, creating the conditions necessary for life and consciousness.

### *Microgravity and the Interplay of Gravity and Cellular Homeostasis*

Experimental findings from studies conducted in microgravity environments further illustrate the interplay between gravitational forces and cellular processes. Torday (2003) observed that cells exposed to microgravity revert to a more primitive state, losing their specialized phenotype. This regression highlights the critical role of gravity in maintaining cellular organization and homeostasis.

In the absence of gravity, the coherence that defines cellular systems becomes disrupted, reflecting the importance of gravitational forces in stabilizing quantum-driven dynamics. These findings suggest that the interplay between gravity and quantum mechanics is integral to cellular function, linking the physical environment to the emergence of complex biological behaviors.

### *Conclusion*

Quantum entanglement serves as a foundational principle for understanding the continuum of consciousness, linking cellular physiology to cosmic processes. Symbiogenesis exemplifies the interconnectedness central to entanglement, while the Diósi-Penrose model provides a theoretical framework for exploring the quantum-classical transition. Experimental evidence from microgravity environments further underscores the importance of gravitational forces in maintaining cellular coherence. Together, these insights position quantum entanglement as a unifying concept that bridges biology, physics, and the study of consciousness.

## Stellar Nucleosynthesis and the Genesis of Consciousness

The origins of life and consciousness are inextricably linked to the elemental processes of the cosmos. Stellar nucleosynthesis, the process by which stars forge heavier elements from lighter ones, serves as the foundation for both the logic of the universe and the building blocks of life. This section explores how stellar processes shaped the evolution of consciousness by providing the raw materials and guiding principles that underpin biological and cognitive systems.

### *Starlight and the Foundations of Life*

Consciousness can be framed as an emergent property of the cosmic processes initiated by stellar nucleosynthesis. Burbidge et al. (1957) demonstrated how elements like hydrogen and helium, created during the Big Bang, were transformed into heavier elements—such as carbon, oxygen, and nitrogen—within the cores of stars. These elements, ejected into the cosmos through supernovae, became the basis for planets, water, and ultimately, life.

Starlight, the byproduct of these elemental reactions, provided the energy necessary for early chemical processes on Earth, such as the formation of lipid micelles and amino acids. These structures formed the precursors to cellular life, setting the stage for the evolution of complex organisms. The elements forged in stars thus laid both the material and energetic foundations for the eventual emergence of life and consciousness.

By framing consciousness as a distant echo of stellar processes, we see its roots not only in biology but also in the fundamental logic of the cosmos.

### *The Sequential Formation of Elements and Cellular Evolution*

The sequential nature of stellar nucleosynthesis parallels the stepwise evolution of cellular systems. Tegmark (2014) proposed that the universe is inherently mathematical, with its structure and dynamics reflecting a logic that extends to living systems. This mathematical order is evident in the periodicity of elemental formation during nucleosynthesis, which mirrors the hierarchical complexity of cellular evolution.

For example, the structural and functional complexity of cells, from lipid bilayers to organelles, reflects the integration of elemental properties. Carbon's versatility in forming stable bonds enabled the creation of complex organic molecules, while oxygen's reactivity facilitated energy transfer within cells through oxidative phosphorylation. These elemental characteristics, shaped by nucleosynthesis, were assimilated into the evolutionary pathways of life.

This integration of cosmic logic into cellular evolution underscores how life is not a random occurrence but a continuation of the universe's ordered processes.

### *From Cosmic Logic to Human Physiology*

The influence of stellar nucleosynthesis extends beyond the molecular building blocks of life to the larger-scale processes that define human physiology. Torday (2024) highlighted how the microcosm of cellular activity reflects the macrocosmic dynamics of the universe. For instance, the circulation of elements like calcium and potassium in the human body mirrors their role in stellar processes, linking physiology to the larger logic of the cosmos.

Calcium, synthesized in supernovae, is critical for cellular signaling and bone formation, while iron, a product of later-stage nucleosynthesis, is essential for oxygen transport in hemoglobin. These elements, integral to human life, trace their origins to the stars, connecting our physiology to the cosmos. Consciousness, as an emergent property of this system, represents the culmination of these cosmic and biological processes.

### *Conclusion*

Stellar nucleosynthesis provides the raw materials and guiding logic that underpin life and consciousness. The elements forged in stars not only form the basis of biological systems but also reflect the mathematical and ordered nature of the universe. By integrating these cosmic principles into the evolution of cellular and human systems, consciousness emerges as a profound expression of the interconnectedness between the microcosm of life and the macrocosm of the cosmos.

## Mathematical Homologies and Evolutionary Processes

The processes of life are deeply interwoven with mathematical principles, reflecting the ordered nature of the universe. From the spiral patterns of galaxies to the organization of cellular systems, mathematical structures such as the Fibonacci sequence and knot theory provide a universal language for describing biological evolution. This section examines how life embodies these inherent mathematical principles, framing evolutionary processes as guided by self-correcting algorithms that respond to environmental challenges.

### *The Fibonacci Sequence and Evolutionary Biology*

The Fibonacci sequence, a series where each number is the sum of the two preceding ones, is ubiquitous in nature. Its manifestations range from the arrangement of leaves on a stem to the spiral patterns of shells. Livio (2002) highlighted the parallels between this mathematical pattern and biological systems, suggesting that the Fibonacci sequence reflects optimization principles in evolutionary biology.

In the context of epigenetic inheritance, the Fibonacci sequence can be seen as a metaphor for how genetic and environmental information is passed across generations. Epigenetic marks, such as DNA methylation, are layered onto genetic material in ways that mirror iterative mathematical algorithms. These marks allow organisms to adapt to their environments while preserving core genetic stability, creating a balance between flexibility and order.

This iterative and predictive quality of epigenetic mechanisms echoes the Fibonacci sequence's role in representing continuity and adaptation, positioning mathematics as an integral framework for understanding evolutionary processes.

## *Rowlands' Rewrite Mathematics and Kauffman's Formal Knot Theory*

Peter Rowlands' Rewrite Mathematics and Louis Kauffman's Formal Knot Theory offer robust frameworks for analyzing the dynamics of cellular and evolutionary processes. Rowlands (2015) proposed that the foundations of mathematics lie in iterative transformations, where systems self-correct and refine their structures over time. This concept aligns closely with biological evolution, where mutations, recombination, and natural selection iteratively shape life's complexity.

Kauffman (2006) extended these ideas into the realm of knot theory, describing cellular processes as analogous to the formation and resolution of knots. During cellular division, for instance, the untangling and replication of DNA can be understood as a topological process where knots are resolved to maintain genomic integrity. Similarly, protein folding involves the creation and stabilization of molecular structures that optimize functionality while minimizing energy expenditure.

By integrating Rewrite Mathematics and knot theory, we gain a deeper understanding of how life organizes itself through mathematical principles, transforming chaos into order and complexity into functionality.

## *Life as a Self-Correcting System*

Life can be framed as a self-correcting system, where mathematical homologies guide adaptive responses to environmental challenges. Rowlands (2015) emphasized that mathematical structures act as attractors, drawing systems toward stability and coherence. In biological systems, this principle is evident in homeostasis, where feedback mechanisms regulate cellular and systemic balance.

For example, the Fibonacci sequence's influence can be seen in the branching patterns of blood vessels, which optimize nutrient distribution. Similarly, knot theory provides a model for understanding the resilience of cellular structures, allowing them to recover from stress or damage. These mathematical homologies act as built-in algorithms that ensure life's continuity, enabling organisms to navigate an ever-changing environment.

The self-correcting nature of life reflects its ability to harness mathematical principles for survival and adaptation, demonstrating the deep connection between biological processes and the underlying logic of the universe.

### *Conclusion*

Mathematical homologies, such as the Fibonacci sequence and knot theory, provide a powerful framework for understanding the dynamics of life. From epigenetic inheritance to protein folding, these principles guide the adaptive processes that define evolution. By integrating concepts from Rewrite Mathematics and formal knot theory, we uncover the self-correcting mechanisms that enable life to transform challenges into opportunities for growth and complexity. This mathematical foundation underscores the profound interconnectedness of biology and the cosmos, positioning evolution as a universal process shaped by the logic of mathematics.

## The Periodic Table and the Cellular Cosmos

The Periodic Table of Elements has long been celebrated as a unifying framework in chemistry, revealing the relationships among atomic structures and their properties. However, its significance extends far beyond the atomic level, offering profound insights into cellular evolution and the underlying principles of life. This section explores the homologies between the Periodic table and cellular systems, drawing parallels between atomic and cellular dynamics to highlight the unity of physical laws.

### *Synchronic and Diachronic Dimensions of the Periodic Table*

The structure of the periodic table encapsulates two essential dimensions: synchronic (horizontal relationships within periods) and diachronic (diagonal relationships across groups). Torday and Rehan (2012) proposed that these dimensions mirror the dual processes of ontogeny (individual development) and phylogeny (evolutionary history) in biological systems. Just as elements within the same period share properties related to their electron configurations, cells within a single developmental stage exhibit common physiological traits. Similarly, the diagonal trends of the periodic table reflect evolutionary trajectories, where shared characteristics evolve incrementally across generations.

This synchronic-diachronic framework offers a compelling lens through which to view cellular evolution. For example, the progression of electron shells within the periodic table parallels the increasing complexity of cellular systems, from unicellular to multicellular organisms. By aligning these dimensions with biological processes, the periodic table serves as a metaphor for the dynamic interplay between development and evolution in living systems.

## *Atom-Cell Homology: Deterministic and Probabilistic Dynamics*

Atoms and cells share a striking homology in their structural and functional dynamics, bridging deterministic and probabilistic principles. Morowitz (2004) highlighted how the deterministic laws of atomic interactions, governed by quantum mechanics, are reflected in the probabilistic behaviors of cellular processes. In both systems, stability arises from the delicate balance of opposing forces—such as the attraction and repulsion of electrons in atoms or the regulation of ion gradients in cells.

The atom-cell analogy is particularly evident in energy transfer mechanisms. In atoms, the quantized energy levels of electrons dictate chemical reactivity, while in cells, energy transfer through molecules like ATP governs metabolic processes. This parallel underscores how deterministic rules at the atomic level scale up to probabilistic behaviors in complex biological systems, uniting the principles of physics and biology.

## *The Pauli Exclusion Principle and Cellular Homeostasis*

The Pauli Exclusion Principle, which states that no two electrons in an atom can occupy the same quantum state simultaneously, plays a foundational role in shaping atomic structure. This principle also finds resonance in cellular systems, where it manifests as a metaphor for the regulatory mechanisms that maintain homeostasis. Morowitz (2004) argued that just as the exclusion principle governs electron configurations, cellular systems employ exclusionary processes to regulate the flow of ions, nutrients, and signals.

For instance, ion channels in cell membranes operate with high specificity, allowing only certain ions to pass under particular conditions. This specificity mirrors the precision of electron arrangements in atoms, reinforcing the unity of physical laws across scales. By applying the principles of quantum mechanics to cellular behavior, the Pauli Exclusion Principle emerges as a universal law connecting the atomic and biological realms.

### *Conclusion*

The periodic table serves as more than a guide to chemical elements; it offers a profound metaphor for understanding cellular evolution and the unity of physical laws. Its synchronic and diachronic dimensions reflect the dual processes of ontogeny and phylogeny, while the atom-cell homology bridges deterministic and probabilistic principles. The Pauli Exclusion Principle further unites atomic and cellular systems, illustrating how the same laws govern the cosmos at all scales. By exploring these parallels, we gain deeper insights into the interconnectedness of the physical and biological worlds, positioning the periodic table as a foundational model for the cellular cosmos.

## Holism, Symbiogenesis, and Consciousness

Consciousness, as a phenomenon that bridges quantum, cellular, and cosmic realms, is best understood through a holistic framework. This section synthesizes the insights from previous discussions, framing consciousness as an emergent property of integrated processes. By examining the interplay of quantum mechanics, cellular evolution, and cosmic dynamics, it positions holism and symbiogenesis as central mechanisms for understanding the unity of life and the nature of consciousness.

### *Consciousness as a Holistic Property*

Consciousness can be framed as a holistic property arising from the integration of quantum phenomena, cellular processes, and cosmic principles. Bohm (1980) introduced the concept of the implicate order, a framework in which all parts of the universe are interconnected and enfolded within a larger whole. This perspective aligns with the quantum cell model, where cellular homeostasis and coherence mirror quantum entanglement and cosmic dynamics.

Torday (2023b) extended this idea, suggesting that consciousness emerges from the seamless integration of these domains. For instance, mitochondrial energy production, governed by quantum tunneling, provides the cellular energy required for cognitive processes, while cosmic principles like entropy regulation shape the fundamental dynamics of life. These interactions create a continuum where consciousness is not confined to the brain but emerges as a property of life's interconnected systems.

By adopting a holistic view, we can reconcile the physical, biological, and experiential aspects of consciousness, positioning it as a unifying phenomenon that reflects the organization of the universe.

## *Symbiogenesis and the Preservation of Biological History*

Symbiogenesis, the process by which distinct organisms integrate to form a new evolutionary entity, offers a powerful mechanism for understanding the emergence of consciousness. Torday (2019) emphasized that symbiogenesis preserves biological history, allowing life to adapt and evolve by assimilating environmental challenges into physiological systems. This historical continuity is reflected in cellular processes, where symbiotic interactions like those between mitochondria and their host cells underpin critical functions such as energy production and signal regulation.

Symbiogenesis also facilitates the integration of diverse systems, enabling the emergence of higher-order properties like consciousness. For example, the nervous system, a product of evolutionary innovation, integrates sensory inputs with memory and cognition, creating a unified experience of awareness. This integrative capacity highlights how symbiogenesis acts as a bridge between biological complexity and the emergence of conscious phenomena.

## *Holism in Biology, Mathematics, and Cosmology*

Holism provides a universal framework for connecting biology, mathematics, and cosmology, emphasizing the unity of life and the cosmos. Wilson (1998) described consilience—the integration of knowledge across disciplines—as essential for understanding complex phenomena. This principle is evident in the interplay of mathematics and biology, where patterns like the Fibonacci sequence and knot theory provide insights into cellular organization and evolution.

Torday (2024) proposed that holism offers a pathway for reconciling seemingly disparate fields, suggesting that consciousness is not an isolated phenomenon but part of a continuum that spans scales and disciplines. For instance, cellular processes governed by physical laws mirror the organization of cosmic systems, linking the microcosm of biology to the macrocosm of the universe. This holistic perspective underscores the interconnectedness of life, positioning consciousness as a property that unites biological and cosmic dynamics.

## *Conclusion*

Consciousness emerges as a holistic property from the integration of quantum, cellular, and cosmic processes. Symbiogenesis preserves biological history and facilitates evolution, enabling the emergence of higher-order phenomena like awareness. Holism, as a framework for understanding the unity of life, connects biology, mathematics, and cosmology, offering profound insights into the interconnected nature of the universe. By embracing this integrative perspective, we can deepen our understanding of consciousness and its role in the fabric of existence.

## References


Bohm, D. (1980). *Wholeness and the implicate order*. Routledge.

Burbidge, E. M., Burbidge, G. R., Fowler, W. A., & Hoyle, F. (1957). Synthesis of the elements in stars. *Reviews of Modern Physics, 29*, 547–650.

Diósi, L. (1987). A universal master equation for the gravitational violation of quantum mechanics. *Physics Letters A, 120*, 377–381.

Kauffman, L. H. (2006). *Formal knot theory*. Dover Publications.

Livio, M. (2002). *The Golden ratio: The story of phi, the world's most astonishing number*. Broadway Books.

Morowitz, H. J. (2004). *The emergence of everything: How the world became complex*. Oxford University Press.

Penrose, R. (2014). On the gravitization of quantum mechanics 1: Quantum state reduction. *Foundations of Physics, 44*, 557–575.

Rowlands, P. (2015). *The foundations of physical law*. World Scientific.

Sagan, L. (1967). On the origin of mitosing cells. *Journal of Theoretical Biology, 14*, 255–274.

Tegmark, M. (2014). *Our mathematical universe: My quest for the ultimate nature of reality*. Knopf.

Torday, J. S. (2003). Parathyroid hormone-related protein is a gravisensor in lung and bone cell biology. *Advances in Space Research, 32*, 1569–1576.

Torday, J. S. (2019). The singularity of nature. *Progress in Biophysics and Molecular Biology, 142*, 23–31.

Torday, J. S. (2023a). Consciousness, embodied quantum entanglement. *Progress in Biophysics and Molecular Biology, 177*, 125–128.

Torday, J. S. (2023b). The holism of cosmology and consciousness. *Progress in Biophysics and Molecular Biology, 177*, 181–184.

Torday, J. S. (2024). Symbiogenesis predicts the monism of the cosmos. *Progress in Biophysics and Molecular Biology, 191*, 58–62.

Torday, J. S., & Rehan, V. K. (2012). *Evolutionary biology, cell-cell signaling and complex disease*. Wiley.

Wilson, E. O. (1998). *Consilience: The Unity of knowledge*. Knopf.

# Chapter 11
# Quantum Entanglement and the Essence of Consciousness

Consciousness, one of the most enigmatic and profound phenomena in nature, has long eluded a comprehensive explanation. Increasingly, researchers are turning to quantum mechanics to illuminate the intricate mechanisms underlying awareness. At the heart of this approach lies quantum entanglement, a phenomenon where particles remain interconnected such that the state of one instantly influences the other, irrespective of distance. This nonlocal property of quantum systems offers a transformative perspective on the essence of consciousness, extending beyond classical neural and biological models.

This chapter delves into the interplay between quantum entanglement and consciousness, positioning it as a unifying principle that bridges the microcosmic world of cellular processes with the macrocosmic dynamics of the universe. By examining the role of entanglement in cellular homeostasis, neural integration, and cognitive emergence, we explore how this quantum phenomenon transcends physical boundaries to facilitate coherent and holistic experiences.

The chapter begins by revisiting the foundations of quantum entanglement, highlighting its experimental evidence and theoretical implications for biology. It then explores the relevance of entanglement to cellular systems, emphasizing its role in maintaining homeostasis and synchronizing complex biological networks. Finally, it considers the broader implications of quantum entanglement for understanding consciousness as an emergent and interconnected property of the cosmos.

Through this lens, consciousness is framed not as an isolated occurrence confined to the human brain but as a fundamental aspect of the universe's quantum fabric. By integrating quantum principles with biological and cognitive sciences, this chapter seeks to provide a deeper understanding of the unity underlying life, awareness, and the cosmos.



J. S. Torday, R. G. Sacco, *Conscious Life*,
https://doi.org/10.1007/978-3-031-80500-4_11

## The Ontological Foundation of Consciousness

The origin and nature of consciousness have perplexed scientists and philosophers for centuries, often framed as a mysterious phenomenon beyond the reach of empirical understanding. However, emerging insights in quantum biology and evolutionary theory offer a compelling perspective: consciousness arises as an emergent property of physiological processes, rooted in both local interactions and nonlocal quantum phenomena. This section explores how the foundational principles of life—physiology, homeostasis, and cosmic interconnectedness—serve as the ontological basis for consciousness.

### *A New Story of Consciousness*

Thomas Berry (1988) called for a "new story," urging humanity to shift its understanding of the universe from a mechanistic to a holistic paradigm. In this new story, the cosmos is not a collection of isolated objects but a "communion of subjects," deeply interconnected at every level. Applying Berry's framework to consciousness, we can begin to see awareness not as a byproduct of neuronal complexity but as a fundamental aspect of the universe's fabric, deeply entwined with life's physiological foundations.

Consciousness, from this perspective, is the cumulative result of interconnected processes that span scales—from the molecular and cellular to the cosmic. This shift from reductionism to holism provides a broader context for understanding consciousness as a continuum rather than a discrete phenomenon.

### *Lipid Micelles and the Quantum Transition to Homeostasis*

The genesis of life marks a critical juncture in the story of consciousness. Deamer (2017) and Torday and Rehan (2012) identified lipid micelles as the precursors to the first living cells, encapsulating critical molecules within semipermeable membranes. These micelles created localized environments where chemical reactions could occur with greater efficiency, enabling the emergence of negative entropy—a key characteristic of living systems.

This transition, governed by quantum principles, established the conditions for homeostasis, where cells maintain internal stability despite external fluctuations. The quantum coherence observed in these early systems allowed for synchronized molecular interactions, ensuring the survival and propagation of primitive cellular life. These foundational processes are the precursors to the physiological complexity that underpins consciousness, linking its origins to the earliest stages of life's evolution.

### *Local and Nonlocal Forces in Consciousness*

The interplay between local and nonlocal phenomena is central to understanding the ontological foundation of consciousness. Einstein's (1961) theories of relativity highlight how local gravitational effects shape the physical environment, influencing everything from the curvature of space-time to the behavior of particles. On a cosmological scale, these gravitational forces interact with quantum principles, creating a dynamic tension between determinism and probability.

Bohm's (1980) concept of the implicate order further bridges these scales, proposing that the universe operates as a unified whole where local and nonlocal phenomena are intertwined. In the context of consciousness, this interplay suggests that awareness arises from the interaction of localized physiological processes—such as neuronal signaling and cellular homeostasis—with the broader quantum dynamics of the cosmos.

### *Conclusion*

Consciousness is not merely a feature of complex neural systems but an emergent property rooted in the physiological and quantum foundations of life. By framing the origin of consciousness through lipid micelles, quantum coherence, and the integration of local and nonlocal forces, we gain a holistic understanding of its nature. This perspective aligns with Berry's call for a "new story," emphasizing interconnectedness as the essence of both life and awareness.

## Symbiogenesis and the Evolution of Homeostasis

Homeostasis, the ability of a system to maintain internal stability despite external fluctuations, is a cornerstone of life. Its emergence and evolution can be traced to symbiogenesis, the process by which disparate biological entities combine to form a unified, adaptive system. By assimilating environmental challenges into cellular physiology, symbiogenesis has not only shaped the development of life but also exemplified the interplay of quantum and classical principles. This section explores how symbiogenesis, rooted in both quantum entanglement and classical biology, facilitated the evolution of homeostatic balance.

### *Symbiogenesis and Cellular Integration*

Symbiogenesis plays a central role in the integration of environmental challenges into cellular physiology. Sagan (1967) proposed that eukaryotic cells emerged through the symbiotic incorporation of prokaryotic organisms, such as the origin of mitochondria from ancient bacteria. This process exemplifies life's ability to transform external threats into physiological advantages, enhancing adaptability and survival.

At the molecular level, the integration of environmental stimuli into cellular function is mediated by quantum entanglement. Torday (2023a) suggested that entangled states within cellular systems enable coherence across vast scales, facilitating the assimilation of local and nonlocal information. For instance, the quantum coherence observed in mitochondrial ATP synthesis underscores how symbiogenesis harnessed quantum principles to maintain energy production and cellular stability.

This ability to integrate and adapt highlights the fundamental role of symbiogenesis in shaping the evolution of homeostasis, demonstrating how life balances the deterministic and probabilistic forces of its environment.

## *Microgravity Studies: The Role of Gravity in Cellular Structure*

Experimental studies conducted in microgravity environments have provided compelling evidence for the role of local gravitational forces in maintaining cellular structure and homeostasis. Torday's (2003) findings revealed that cells exposed to microgravity lose their phenotypic identity, reverting to a more primitive state. This phenomenon underscores the reliance of cellular systems on local gravitational forces to maintain their organization and function.

In microgravity, the absence of directional cues disrupts the cytoskeletal structure, leading to impaired cellular signaling and a loss of differentiation. These findings suggest that local gravity serves as a stabilizing force, anchoring cells within their homeostatic framework. The reversion of cells to an undifferentiated state in microgravity mirrors the flexibility and adaptability inherent to symbiogenesis, where cellular systems reconfigure themselves in response to environmental changes.

By highlighting the interplay between gravitational forces and cellular homeostasis, these studies provide a deeper understanding of how life negotiates the boundary between quantum uncertainty and classical stability.

## *The Unity of Quantum and Classical Principles in Homeostasis*

Symbiogenesis exemplifies the seamless integration of quantum and classical principles in maintaining homeostasis. Torday (2022) framed this process as a dynamic equilibrium where cells assimilate both local (gravitational and environmental) and nonlocal (quantum) information. For example, the regulation of ion channels, critical for cellular signaling, involves quantum tunneling phenomena that operate alongside classical diffusion mechanisms.

This duality reflects the holistic nature of life, where quantum mechanics provides the coherence necessary for molecular interactions, and classical principles ensure structural and functional stability. Symbiogenesis, by assimilating diverse inputs into a unified system, embodies this interplay, enabling life to adapt and evolve in complex environments.

### *Conclusion*

Symbiogenesis serves as a foundational mechanism for the evolution of homeostasis, bridging the quantum and classical realms. By integrating environmental challenges into cellular physiology, it exemplifies the adaptability and resilience of life. Experimental findings from microgravity studies further highlight the role of local gravitational forces in maintaining cellular structure and function. Together, these insights position symbiogenesis as a unifying process that connects the molecular, cellular, and cosmic dimensions of life.

## Local and Nonlocal Consciousness

Consciousness exists at the intersection of local physiological processes and nonlocal cosmic principles, forming a continuum that bridges the microcosmic and macrocosmic realms. This duality reflects the ability of consciousness to integrate immediate sensory experiences with universal, quantum-level phenomena. By examining the interplay between quantum entanglement, epigenetic inheritance, and experimental evidence, this section highlights how local and nonlocal processes collectively shape awareness and perception.

### *Quantum Entanglement and the Integration of Local and Cosmic States*

Quantum entanglement provides a powerful framework for understanding the integration of local physiological states with nonlocal cosmological forces. Genovese and Gramegna (2019) demonstrated how entangled particles remain interconnected regardless of spatial separation, enabling instantaneous communication. In biological systems, this phenomenon may underpin cellular coherence, where molecular processes operate in synchrony despite spatial or temporal constraints.

Torday (2023b) extended this concept to consciousness, suggesting that quantum entanglement allows the brain and body to align local physiological states with broader cosmological dynamics. For example, mitochondrial energy production involves quantum tunneling, linking cellular function to quantum principles. Similarly, neural networks may utilize entanglement to synchronize activity across different brain regions, facilitating holistic perception and decision-making.

This interconnectedness positions consciousness as both a localized phenomenon rooted in physiology and a nonlocal property influenced by quantum and cosmic forces, reflecting the unity of life and the universe.

## *Body Memory and Epigenetic Inheritance*

The phenomenon of "body memory" further illustrates the duality of consciousness, connecting individual experiences with environmental and cosmological influences. King and Skinner (2020) highlighted how epigenetic inheritance preserves environmental information across generations, encoding adaptive responses in the genome. This process reflects a form of memory that transcends individual lifespans, linking present states to ancestral and cosmic histories.

Pert (1997) explored the role of neuropeptides in body memory, demonstrating how emotional and sensory experiences are stored within the body's physiological systems. These findings suggest that consciousness is not confined to the brain but distributed throughout the body, integrating local experiences with nonlocal influences.

This distributed nature of consciousness highlights its dual role as both a repository of immediate experiences and a medium for connecting with universal patterns, underscoring the interplay between local and nonlocal processes.

## *Experimental Evidence: The Dual Nature of Perception*

Experimental findings such as the Libet Experiment and the double-slit experiment offer compelling evidence for the dual nature of consciousness and perception. Libet (1985) showed that unconscious neural activity precedes conscious decision-making, suggesting that perception and awareness are influenced by both deterministic and probabilistic factors. This aligns with the notion of quantum superposition, where multiple potential states exist simultaneously until observed.

The double-slit experiment (Young, 1804) further supports this perspective, revealing how the act of observation collapses quantum wave functions into distinct particle states. This experiment underscores the role of consciousness in shaping reality, bridging quantum uncertainty with classical determinism.

Together, these findings highlight the dual nature of consciousness, where local physiological processes and nonlocal quantum states converge to create coherent perception and awareness.

### *Conclusion*

Consciousness operates as a dynamic interplay between local and nonlocal phenomena, integrating immediate physiological states with cosmic and quantum influences. Quantum entanglement provides the coherence necessary for this integration, while body memory and epigenetic inheritance link individual experiences to broader environmental and universal patterns. Experimental evidence such as the Libet and double-slit experiments underscores the dual nature of perception, positioning consciousness as both a localized and nonlocalized phenomenon. This duality reflects the fundamental interconnectedness of life and the cosmos, offering a holistic perspective on the nature of awareness.

## Physiological Integration and the Continuum of Consciousness

Consciousness, often perceived as a uniquely human trait, is increasingly recognized as a continuum that spans species and scales, deeply embedded in physiological processes. This section explores how epigenetics, the endocrine system, and comparative consciousness studies reveal the intricate mechanisms that integrate environmental, emotional, and cognitive inputs into a unified conscious experience. By framing consciousness as a physiological continuum, this perspective emphasizes its roots in cellular and systemic processes.

### *Epigenetics and Cellular Memory as Foundations of Consciousness*

Epigenetics serves as a critical mediator of environmental adaptation, linking external stimuli to heritable physiological and behavioral changes. Ben Maamar et al. (2021) demonstrated how epigenetic modifications, such as DNA methylation and histone acetylation, preserve adaptive responses across generations. This process reflects a form of cellular memory, enabling organisms to encode environmental challenges into their genomic architecture.

Cellular memory extends beyond genetic material, capturing patterns of environmental interaction that inform future responses. In the context of consciousness, these epigenetic and cellular mechanisms act as repositories for experience, allowing the inheritance of traits that contribute to awareness and decision-making. For instance, stress-induced epigenetic changes in parental generations can influence offspring behavior, demonstrating the continuity of adaptive consciousness traits.

By anchoring consciousness in the molecular processes of epigenetics and cellular memory, we begin to see it not as a discrete phenomenon but as an evolving, integrative property of life.

## *The Endocrine System: Molecules of Emotion*

The endocrine system plays a central role in transmitting emotional and environmental information throughout the body, linking physiological states to conscious awareness. Pert (1997) introduced the "molecules of emotion" hypothesis, highlighting how neuropeptides and hormones act as biochemical messengers that encode emotional states into bodily functions. For example, the release of cortisol in response to stress connects the external environment to internal physiological adaptations, influencing perception and behavior.

This biochemical network underscores the interconnectedness of body and mind, where emotional states influence not only mental awareness but also physical health. The endocrine system's role in mediating these interactions illustrates the physiological basis of consciousness, where hormonal signals integrate external and internal stimuli into a cohesive experience.

By framing the endocrine system as a dynamic communication network, this perspective reinforces the idea of consciousness as a distributed property, deeply rooted in physiological processes.

## *Comparative Consciousness: The Cambridge Declaration*

The Cambridge Declaration on Consciousness (Low, 2012) marked a milestone in the scientific acknowledgment of consciousness as a continuum across species. The declaration affirmed that nonhuman animals, including birds, mammals, and even cephalopods, possess neurological substrates capable of supporting conscious experience. This recognition expands our understanding of consciousness beyond the human brain, situating it within a broader evolutionary and physiological context.

The declaration highlights shared mechanisms, such as neural plasticity and sensory integration, that underpin consciousness across species. By emphasizing the continuity of these processes, it challenges the anthropocentric view of consciousness, instead proposing it as a universal phenomenon rooted in biology.

This comparative approach underscores the importance of studying consciousness not only as a human trait but also as a property inherent to life itself, shaped by shared physiological mechanisms.

## *Conclusion*

Consciousness emerges from the integration of physiological mechanisms, spanning molecular, systemic, and evolutionary scales. Epigenetics and cellular memory provide a framework for adapting and inheriting traits that contribute to awareness, while the endocrine system acts as a biochemical network for encoding emotional and environmental information. The Cambridge Declaration on Consciousness further underscores the continuity of consciousness across species, linking it to universal physiological principles. By framing consciousness as a continuum rooted in biology, this perspective offers a holistic understanding of its origins and functions.

## Implications for the Philosophy of Mind and Future Directions

The exploration of consciousness as an emergent property of life, rooted in physiology and informed by quantum mechanics, offers profound implications for the philosophy of mind and interdisciplinary research. This section synthesizes insights from prior discussions, integrating philosophical frameworks, quantum biology, and physiological principles to propose future directions for understanding consciousness as both a biological and cosmological phenomenon.

### *Reframing Chalmers' Hard Problem and Clark's Extended Mind*

David Chalmers (1995) posed the "hard problem" of consciousness, questioning how subjective experience arises from objective physical processes. By situating consciousness within the framework of embodied physiology, this dilemma can be reframed: rather than a gap to be bridged, subjective experience reflects the integration of cellular, quantum, and systemic processes. For instance, the interplay between cellular homeostasis and quantum coherence offers a biological basis for understanding how awareness emerges from the interaction of localized and nonlocalized phenomena.

Andy Clark's (1998) "extended mind" hypothesis further enriches this perspective, proposing that consciousness is not confined to the brain but extends into the body and environment. This aligns with the endocrine system's role in encoding emotional states and the cellular memory mechanisms discussed earlier. Together, Chalmers' and Clark's ideas point to a holistic understanding of consciousness as a distributed property that transcends traditional boundaries between mind, body, and environment.

### *Integrating Quantum Biology into Neuroscience*

The integration of quantum biology into neuroscience offers a revolutionary approach to studying consciousness. Quantum coherence, entanglement, and tunneling provide mechanisms for understanding phenomena such as neural synchronization and the emergence of awareness. Torday (2022) emphasized the importance of quantum processes in cellular function, suggesting that these principles extend to neural networks, where quantum phenomena facilitate communication and integration across the brain.

By incorporating quantum mechanics into the study of the brain, neuroscience can move beyond reductionist models, embracing a more holistic perspective that accounts for the dynamic interplay between physiology and quantum physics. This integrative approach offers the potential to resolve long-standing questions about the nature of consciousness, positioning it as a phenomenon that reflects the unity of physical and biological systems.

## *Consciousness as an Emergent Property of Life*

Consciousness, as an emergent property of life, reflects the unity of physiological, quantum, and cosmological processes. Bohm's (1980) implicate order offers a framework for understanding how these levels interact, proposing that the universe operates as a coherent whole where local and nonlocal phenomena are seamlessly integrated. This perspective aligns with Berry's (1988) call for a "communion of subjects," emphasizing interconnectedness as the essence of existence.

The implications of this holistic view extend beyond biology and neuroscience, offering insights into ontological and epistemological questions. By framing consciousness as a product of life's evolution, shaped by both quantum mechanics and environmental adaptation, this perspective challenges traditional dichotomies between materialism and idealism, proposing a unified framework that bridges these paradigms.

## *Future Directions in Consciousness Research*

The study of consciousness demands interdisciplinary approaches that integrate biology, physics, neuroscience, and philosophy. Future research could explore:

- The role of quantum phenomena in neural processes, using advanced imaging techniques to identify signatures of coherence and entanglement in the brain.
- The evolutionary origins of consciousness, linking cellular memory and epigenetics to the emergence of cognitive complexity.
- Philosophical frameworks that reconcile subjective and objective dimensions of consciousness, drawing on insights from process philosophy and systems theory.

By embracing these directions, researchers can deepen their understanding of consciousness as a phenomenon that reflects the interconnectedness of life and the cosmos.

## *Conclusion*

The exploration of consciousness through the lens of physiology, quantum mechanics, and philosophy reveals its nature as an emergent property of life, deeply rooted in the interconnected dynamics of the universe. Reframing Chalmers' hard problem and Clark's extended mind hypothesis within this holistic framework highlights the potential for interdisciplinary research to resolve foundational questions about awareness and existence. By integrating quantum biology into neuroscience and embracing holistic approaches, the study of consciousness can move toward a unified understanding of life and the cosmos.

## References

Ben Maamar, M., Sadler-Riggleman, I., Beck, D., & Skinner, M. K. (2021). Epigenetic transgenerational inheritance of paternal stress-induced pathologies. *Scientific Reports, 11*(1), 1463.
Berry, T. (1988). *The dream of the earth*. Sierra Club Books.
Bohm, D. (1980). *Wholeness and the implicate order*. Routledge.
Chalmers, D. J. (1995). Facing up to the problem of consciousness. *Journal of Consciousness Studies, 2*(3), 200–219.
Clark, A. (1998). *Being there: Putting brain, body, and world together again*. MIT Press.
Deamer, D. (2017). The role of lipid membranes in life's origin. *Life, 7*(1), 5.
Einstein, A. (1961). *Relativity: The special and the general theory*. Crown Publishers.
Genovese, M., & Gramegna, M. (2019). Quantum entanglement in biological systems: A conceptual framework. *Nature Physics, 15*(5), 481–489.
King, S. E., & Skinner, M. K. (2020). Epigenetic transgenerational inheritance of adaptive responses. *Nature Reviews Genetics, 21*(1), 1–13.
Libet, B. (1985). Unconscious cerebral initiative and the role of conscious will in voluntary action. *Behavioral and Brain Sciences, 8*(4), 529–539.
Low, P. (2012). The Cambridge declaration on consciousness. *FENS Forum, 2012*.
Pert, C. (1997). *Molecules of emotion: The science behind mind-body medicine*. Scribner.
Sagan, L. (1967). On the origin of mitosing cells. *Journal of Theoretical Biology, 14*(3), 255–274.
Torday, J. S. (2003). Parathyroid hormone-related protein is a gravisensor in lung and bone cell biology. *Advances in Space Research, 32*, 1569–1576.
Torday, J. S. (2022). Quantum mechanics predicts evolutionary biology. *Progress in Biophysics and Molecular Biology, 135*, 11–15.
Torday, J. S. (2023a). Consciousness, embodied quantum entanglement. *Progress in Biophysics and Molecular Biology, 177*, 125–128.
Torday, J. S. (2023b). The holism of cosmology and consciousness. *Progress in Biophysics and Molecular Biology, 177*, 181–184.
Torday, J. S., & Rehan, V. K. (2012). *Evolutionary biology, cell-cell communication, and complex disease*. Wiley.
Young, T. (1804). Experiments and calculations relative to physical optics. *Philosophical Transactions of the Royal Society of London, 94*, 1–16.

# Chapter 12
# Dual Time Perspectives in Cellular Consciousness

Time, the ever-present yet elusive dimension of existence, shapes every aspect of life, from the movement of celestial bodies to the biochemical rhythms of a single cell. In the context of consciousness, time reveals itself in two distinct yet interconnected forms: the synchronic, representing the immediate and simultaneous, and the diachronic, embodying the historical and sequential. These dual perspectives of time provide a framework for understanding the evolution of cellular consciousness, where the present moment intersects with the legacy of evolutionary memory.

This chapter explores how cellular processes reflect and integrate these two time perspectives. Synchronicity emerges in real-time cellular interactions, such as ion channel regulation and metabolic pathways, while diachronicity manifests in the transmission of genetic information and the evolutionary adaptation to environmental challenges. Together, these temporal frameworks form the basis for cellular homeostasis, memory, and ultimately, the emergence of consciousness.

Drawing from concepts in evolutionary biology, quantum mechanics, and systems theory, this chapter examines the profound interplay between synchronic and diachronic time within the cellular context. From the dynamic equilibrium of molecular interactions to the heritable patterns encoded in DNA, time serves as both a constraint and a guide for life's unfolding complexity.

By delving into the dual time perspectives of cellular consciousness, this chapter aims to uncover the mechanisms by which cells integrate the immediate and the historical, offering insights into the continuity of life and the emergence of awareness. Through this lens, we gain a deeper appreciation of how time, both synchronic and diachronic, serves as a unifying principle connecting the microcosm of the cell to the macrocosm of the universe.



J. S. Torday, R. G. Sacco, *Conscious Life*,
https://doi.org/10.1007/978-3-031-80500-4_12

## Defining Cellular Holism

Cellular holism offers a profound framework for understanding the dual temporal dynamics—synchronic and diachronic—through which cells adapt, evolve, and sustain life. As the fundamental building blocks of all living organisms, cells integrate immediate environmental stimuli with the evolutionary memory encoded in their genetic and epigenetic structures. This integration provides a pathway to understanding consciousness, rooted in the dynamic interplay of synchronic adaptation (moment-to-moment) and diachronic processes (across evolutionary time). Cellular holism, with homeostasis as its guiding principle, bridges the temporal and structural dimensions of life.

### *Synchronic and Diachronic Adaptation: A Cellular Perspective*

At the heart of cellular holism lies the ability of cells to adapt synchronically and diachronically. Synchronic adaptation refers to real-time cellular responses, such as ion channel regulation or metabolic shifts, allowing cells to maintain equilibrium within rapidly changing environments. For instance, the regulation of calcium ions, a critical signaling mechanism, exemplifies how cells achieve immediate homeostasis in response to external stimuli (Torday, 2023). These moment-to-moment adjustments are essential for sustaining life in the face of constant environmental fluctuations.

Diachronic adaptation, in contrast, unfolds over evolutionary timescales, as cells encode adaptive responses into their genetic and epigenetic frameworks. This historical perspective is evident in the evolution of organ systems, where cellular differentiation and specialization mirror the long-term adjustments made in response to environmental pressures. By integrating both temporal dimensions, cells function as holistic entities that embody the continuity of life.

### *The Role of Gravitational Forces in Cellular Differentiation*

Gravitational forces play a pivotal role in cellular differentiation, influencing both structure and function. Microgravity experiments have demonstrated that the absence of gravitational cues disrupts cellular organization, leading to the loss of phenotypic identity and a reversion to a more primitive, undifferentiated state (Torday, 2003). This phenomenon underscores the reliance of cellular systems on local gravitational forces for maintaining their specialized functions.

Deamer (2017) highlighted how lipid micelle formation, driven by physical forces such as gravity, served as the foundational step in the emergence of cellular life. These micelles provided the structural integrity needed for biochemical reactions to occur within a controlled environment, setting the stage for cellular differentiation. Gravitational forces, therefore, not only shaped the physical origins of life but also continue to influence cellular adaptation and specialization across time.

### *Cellular Holism and Bohm's Implicate and Explicate Orders*

David Bohm's (1980) concept of the Implicate and Explicate Orders provides a philosophical framework for understanding cellular holism. The Implicate Order represents a deeper, interconnected reality where all phenomena are unified, while the Explicate Order manifests as the observable, differentiated structures of the physical world. Cellular processes reflect this duality, with synchronic adaptations mirroring the Explicate Order and diachronic processes resonating with the Implicate Order.

Cells, as holistic entities, embody this duality by integrating immediate functional responses with evolutionary memory. For example, the homeostatic mechanisms that maintain cellular equilibrium operate within the Explicate Order, while the long-term adaptations encoded in genetic material reflect the Implicate Order's continuity. This framework positions cells as microcosms of the broader universe, bridging the divide between the temporal and the eternal.

### *Conclusion*

The concept of cellular holism offers a unifying perspective on the synchronic and diachronic dimensions of life, framing cells as entities that seamlessly integrate real-time adaptation with evolutionary history. Gravitational forces and cellular differentiation highlight the importance of local physical cues, while Bohm's Implicate and Explicate Orders provide a philosophical lens for understanding the interconnectedness of cellular processes. Together, these insights illuminate how cells serve as the foundation for consciousness, embodying the dynamic interplay between the immediate and the eternal.

## Evolutionary Transitions and Cellular Homeostasis

Evolutionary transitions, marked by significant adaptations to environmental challenges, reveal the diachronic role of cells in shaping the trajectory of life. Cellular mechanisms, such as homeostasis, symbiogenesis, and epigenetic inheritance, form the foundation for these transitions, enabling organisms to internalize environmental pressures and translate them into adaptive responses. This section explores key evolutionary shifts, emphasizing the cellular processes that underpin life's resilience and continuity.

## *From Aquatic to Terrestrial Life: Cellular Adaptations to Gravity*

The transition from aquatic to terrestrial environments stands as one of the most profound evolutionary milestones. This shift required extensive cellular and systemic adaptations to overcome challenges posed by gravity, desiccation, and oxygen availability. Lung alveolarization, the development of small air sacs that maximize oxygen exchange, exemplifies how cells adapted to terrestrial life. Clack (2012) highlighted how early tetrapods evolved lungs with increased surface area, enabling efficient gas exchange to meet the demands of life on land.

Similarly, skeletal stiffening, driven by osteoblast activity, provided structural support in the absence of aquatic buoyancy. Rubin et al. (2004) demonstrated that skeletal cells responded to gravitational forces by remodeling bone tissue, optimizing its density and strength. These adaptations reflect the ability of cellular homeostasis to regulate physiological systems in response to external pressures, illustrating the diachronic integration of environmental challenges into biological structures.

## *Symbiogenesis: Integrating Environmental Challenges*

Symbiogenesis, first proposed by Sagan (1967), serves as a critical mechanism for integrating environmental challenges into cellular physiology. This process involves the incorporation of one organism into another, creating a symbiotic relationship that drives evolutionary innovation. For example, the origin of mitochondria as energy-producing organelles within eukaryotic cells exemplifies how symbiogenesis transformed a survival challenge—oxygen toxicity—into a physiological advantage.

Through symbiogenesis, cells internalize external stressors, transforming them into functional components of their physiology. Torday (2024) framed this process as a continuous negotiation between cells and their environments, where survival hinges on the ability to assimilate and repurpose external factors. This mechanism not only facilitates immediate adaptation but also lays the groundwork for long-term evolutionary transitions, positioning cells as dynamic entities capable of integrating environmental inputs across scales.

## *Cellular Memory and Epigenetic Inheritance*

Cellular memory, encoded through epigenetic mechanisms, enables organisms to adapt to environmental changes across generations. Ben Maamar et al. (2021) highlighted how DNA methylation and histone modifications preserve adaptive responses, allowing offspring to inherit traits that enhance their survival. This form of epigenetic inheritance extends the capacity of cells to respond diachronically, linking past environmental challenges to present and future adaptations.

For example, stress-induced epigenetic changes in parental generations can influence offspring physiology and behavior, demonstrating the role of cellular memory in shaping evolutionary trajectories. This process reflects the diachronic nature of cellular homeostasis, where adaptations are encoded and transmitted across time, ensuring the resilience of life in changing environments.

### *Conclusion*

The diachronic role of cells in evolutionary transitions underscores their centrality to life's adaptability and continuity. From the transition to terrestrial life to the mechanisms of symbiogenesis and epigenetic inheritance, cells integrate environmental challenges into their physiology, driving innovation and resilience. By framing these transitions within the context of cellular homeostasis, this section highlights the dynamic interplay between immediate adaptation and long-term evolution, positioning cells as the architects of life's enduring complexity.

## Cellular Holism and the Philosophy of Consciousness

Cellular holism extends beyond biology into the realms of philosophy, art, and epistemology, offering profound insights into the nature of life and consciousness. By framing the unicell as a metaphor for wholeness, cellular holism captures the intricate interplay between individuality and interconnectedness, mirroring broader philosophical principles. This section delves into the philosophical implications of cellular holism, highlighting its connections to artistic expression and foundational concepts in logic and uncertainty.

### *The Unicell as a Metaphor for Wholeness*

The unicellular organism serves as a profound metaphor for wholeness, embodying the principles of self-containment and dynamic equilibrium. Its striving for homeostasis, a state of internal balance amidst external fluctuations, resonates with the Zen koan of "one hand clapping," symbolizing the pursuit of harmony in apparent contradiction (Torday, 2021). The functional architecture of the unicell, akin to a Möbius strip, integrates memory and adaptability, allowing it to respond to environmental challenges while maintaining its structural integrity.

This Möbius strip-like quality highlights the unity of opposites within cellular processes, where inside and outside, past and future, converge seamlessly. It is within this framework of holistic integration that the unicell reflects a deeper philosophical truth: the inherent interconnectedness of all life forms. By striving for homeostasis, cells exemplify the dynamic tension between individuality and universality, a concept central to understanding consciousness.

## *Cellular Holism and Artistic Expression*

Artistic expression provides a compelling parallel to cellular holism, illustrating how structural and functional principles in biology resonate with the creative processes of music, literature, and visual arts. Saunders (2022) explored how evolutionary holism is reflected in the recursive patterns of music, the narrative arcs of literature, and the symmetry of visual arts. For instance, the repetitive yet evolving structures in classical symphonies mirror the iterative cycles of cellular adaptation and evolution.

Similarly, the interconnected themes in great literary works, where individual characters and broader narratives intertwine, parallel the interplay of synchronic and diachronic processes within cells. Visual arts, particularly those that emphasize balance and asymmetry, reflect the cellular struggle for homeostasis in dynamic environments. These artistic analogies underscore the universality of holism, bridging biological and creative domains to offer a richer understanding of life and consciousness.

## *Philosophical Frameworks and Cellular Holism*

Cellular holism aligns with several foundational philosophical principles, such as Gödel's Incompleteness Theorem and Heisenberg's Uncertainty Principle. Gödel's theorem, which posits that no formal system can be both complete and consistent, parallels the inherent openness of cellular systems, where adaptation and innovation ensure survival but preclude static perfection (Torday & Rehan, 2012). Similarly, Heisenberg's principle, which highlights the limits of precision in measuring quantum systems, resonates with the dynamic unpredictability of cellular processes, emphasizing their dual role as observer and observed.

Bohm's (1980) concept of the Implicate and Explicate Orders further illuminates cellular holism. Cells operate within the Explicate Order of observable phenomena while simultaneously reflecting the Implicate Order of interconnected reality. This duality positions cells as microcosmic representations of the universe, embodying the philosophical notion that life and consciousness are emergent properties of holistic systems.

### *Conclusion*

The philosophy of cellular holism reveals the profound interconnectedness of life, where the unicell serves as a metaphor for wholeness, and artistic expression echoes biological principles. By situating cellular processes within broader philosophical frameworks, such as Gödel's Incompleteness Theorem and Bohm's Implicate Order, cellular holism bridges biology, art, and philosophy. This integrative perspective not only deepens our understanding of consciousness but also underscores its universal nature, rooted in the dynamic interplay of individuality and interconnectedness.

## Cosmic and Cellular Integration

The interconnectedness of cellular processes and cosmic dynamics underscores the holistic nature of life and its origins. From the emergence of lipid micelles as precursors to cellular life to the synchronic and diachronic properties of the periodic table, the interplay between cosmic and cellular processes reveals a profound unity. This section explores how these dynamics converge, framing consciousness as an emergent property of this integration.

### *Lipid Micelles and Cosmic Origins*

The origins of life are intrinsically tied to the cosmos. Lipid micelles, often considered the precursors to cellular life, emerged from materials forged in the heart of stars and distributed through cosmic events. McGuire et al. (2021) demonstrated that pulsar-generated lipids and other organic molecules formed in interstellar environments provide the fundamental building blocks of life. These materials, carried to Earth by meteorites and comets, laid the groundwork for the formation of micelles, which serve as self-assembling, semipermeable structures capable of encapsulating biochemical reactions.

The process of micelle formation reflects the principles of energy optimization and structural stability, linking physical and chemical processes to the genesis of life. This cosmic connection establishes a continuum between the material origins of the universe and the emergence of biological systems, positioning micelles as a bridge between the inanimate and animate.

## *Cellular Holism and the Periodic Table*

The synchronic (spatial) and diachronic (temporal) properties of the periodic table mirror the dynamics of cellular evolution, highlighting the parallels between atomic and biological systems. Widrow et al. (2014) explored how the arrangement of elements in the periodic table reflects both immediate interactions (synchronicity) and historical processes (diachronicity), such as stellar nucleosynthesis and radioactive decay. Similarly, cellular evolution involves synchronic adaptations to immediate environmental conditions and diachronic integration of genetic and epigenetic changes across generations.

For example, cellular processes like ion exchange and molecular signaling rely on elements such as calcium, potassium, and sodium, whose properties are shaped by their positions on the periodic table. These elements enable homeostasis, a fundamental feature of cellular life. The periodicity of elements, characterized by predictable properties and behaviors, resonates with the repetitive yet dynamic patterns observed in cellular evolution, emphasizing the unity of physical laws across scales.

## *Consciousness as an Emergent Integration*

Consciousness arises from the integration of cellular and cosmic dynamics, where local processes interact with nonlocal forces to create a unified system. Bohm's (1980) concept of the Implicate Order, which describes an underlying interconnected reality, provides a framework for understanding how cellular processes reflect cosmic principles. Cellular holism, with its integration of synchronic and diachronic adaptations, mirrors the interconnectedness of the universe, where local phenomena are influenced by global dynamics.

Torday (2023) emphasized that consciousness emerges from the interplay of these forces, where quantum entanglement, gravitational interactions, and biochemical processes converge. This perspective positions consciousness not as an isolated phenomenon but as a manifestation of the universe's inherent unity, reflecting the dynamic relationships between the microcosm of the cell and the macrocosm of the cosmos.

## *Conclusion*

The integration of cellular and cosmic dynamics highlights the holistic nature of life and consciousness. From the emergence of lipid micelles to the parallels between the periodic table and cellular evolution, these processes illustrate the profound unity of physical and biological systems. Consciousness, as an emergent property of this integration, bridges local and nonlocal phenomena, providing a framework for understanding life as an interconnected whole.

## Implications for Understanding Consciousness

The exploration of cellular holism provides a transformative perspective on the nature of consciousness, offering insights that bridge biology, quantum mechanics, and philosophy. By integrating synchronic cellular processes with diachronic evolutionary mechanisms, cellular holism serves as a foundational framework for addressing the complexities of consciousness and the mind-body problem. This section synthesizes these insights, discussing their implications for scientific research, philosophical inquiry, and practical applications.

### *Consciousness as a Product of Synchronic and Diachronic Integration*

Consciousness emerges from the interplay between synchronic (real-time) cellular processes and diachronic (historical) evolutionary mechanisms. Synchronic processes, such as neural signaling and metabolic homeostasis, maintain the dynamic balance necessary for life, while diachronic mechanisms encode adaptive responses through genetic and epigenetic modifications. Together, these processes create a continuum that links the present moment to evolutionary history.

Tononi's (2004) Integrated Information Theory (IIT) posits that consciousness arises from the integration of information within a system, aligning closely with the concept of cellular holism. Cells, as units of integration, process and store information across time, reflecting the dual temporal dimensions of consciousness. This perspective offers a novel approach to the mind-body problem, framing consciousness not as a separate entity but as an emergent property of biological systems.

### *Applications in Medicine, Neuroscience, and Philosophy*

The principles of cellular holism have profound implications for medicine, neuroscience, and philosophy. In medicine, understanding consciousness as an emergent property of cellular processes opens new avenues for treating conditions such as neurodegenerative diseases and mental health disorders. By targeting cellular homeostasis and epigenetic pathways, therapeutic strategies can address the root causes of dysfunction, rather than merely alleviating symptoms.

In neuroscience, cellular holism emphasizes the importance of integrating biological and quantum perspectives. Margulis (2001) highlighted how symbiogenesis and cellular adaptation drive evolutionary innovation, suggesting that similar integrative processes underlie neural function. Quantum effects, such as entanglement and superposition, may play a role in synaptic connectivity and information processing, providing a deeper understanding of neural dynamics.

Philosophically, cellular holism challenges reductionist approaches to consciousness, advocating for a more integrative perspective. Torday (2023) argued that the unity of life, consciousness, and the cosmos reflects a fundamental interconnectedness that transcends disciplinary boundaries. This perspective encourages a holistic approach to philosophical inquiries into the nature of mind, self, and existence.

## *The Unity of Life, Consciousness, and the Cosmos*

Cellular holism underscores the unity of life, consciousness, and the cosmos, emphasizing the interconnectedness of all levels of existence. The dynamic equilibrium maintained by cells mirrors the balance observed in cosmic systems, such as the fine-tuning of physical constants that allow life to emerge (Torday, 2024). This holistic view positions consciousness as a universal phenomenon, rooted in the same principles that govern the universe.

Interdisciplinary research is crucial for advancing our understanding of consciousness. By integrating insights from biology, physics, and philosophy, researchers can develop a comprehensive framework that unites diverse perspectives. This approach not only deepens our understanding of consciousness but also fosters collaboration across fields, driving innovation and discovery.

## *Conclusion*

The implications of cellular holism extend far beyond biology, offering a unifying framework for understanding consciousness, life, and the cosmos. By linking synchronic and diachronic processes, cellular holism provides a new lens for addressing the mind-body problem and advancing research in medicine, neuroscience, and philosophy. As a model of interconnectedness, cellular holism emphasizes the need for interdisciplinary approaches, highlighting the unity of existence and the profound connections between cells, consciousness, and the universe.

## References

Ben Maamar, M., Sadler-Riggleman, I., Beck, D., & Skinner, M. K. (2021). Epigenetic transgenerational inheritance of paternal stress-induced pathologies. *Scientific Reports, 11*(1), 1463.
Bohm, D. (1980). *Wholeness and the implicate order*. Routledge.
Clack, J. A. (2012). *Gaining ground: The origin and evolution of Tetrapods*. Indiana University Press.
Deamer, D. (2017). The role of lipid membranes in life's origin. *Life, 7*(1), 5.
Margulis, L. (2001). *Symbiotic planet: A new look at evolution*. Basic Books.
McGuire, B. A., Brogan, C. L., Hunter, T. R., et al. (2021). Detection of interstellar propargylimine. *Science Advances, 7*(6), eabd7270.

Rubin, C. T., Judex, S., & Qin, Y. X. (2004). Skeletal adaptation to mechanical stimuli in the absence of weight bearing. *Journal of Bone and Mineral Research, 19*(9), 1458–1465.

Sagan, L. (1967). On the origin of mitosing cells. *Journal of Theoretical Biology, 14*(3), 255–274.

Saunders, S. (2022). Evolutionary holism in art and science. *Journal of Aesthetic Science, 14*(2), 105–123.

Tononi, G. (2004). An information integration theory of consciousness. *BMC Neuroscience, 5*, 42.

Torday, J. S. (2003). Parathyroid hormone-related protein is a gravisensor in lung and bone cell biology. *Advances in Space Research, 32*, 1569–1576.

Torday, J. S. (2021). Life is a Möbius strip. *Progress in Biophysics and Molecular Biology, 167*, 41–45.

Torday, J. S. (2023). Consciousness, embodied quantum entanglement. *Progress in Biophysics and Molecular Biology, 177*, 125–128.

Torday, J. S. (2024). The synchronic, diachronic cell as the holism of consciousness. *Progress in Biophysics and Molecular Biology, 188*, 19–23.

Torday, J. S., & Rehan, V. K. (2012). *Evolutionary biology, cell-cell communication, and complex disease*. Wiley.

Widrow, L. M., Dubinski, J., & Kim, S. (2014). The cosmic origin of the periodic table. *Astronomical Review, 9*(1), 34–49.

# Part IV
# Evolutionary Patterns: From Biology to Astronomy

# Chapter 13
# The Story of Vertebrate Evolution

The evolution of vertebrates is one of the most remarkable narratives in the history of life. It is a story of innovation, adaptation, and resilience, tracing the transition from simple aquatic organisms to the complex, highly specialized vertebrates that dominate today's ecosystems. This chapter delves into the evolutionary trajectory of vertebrates, revealing how environmental pressures, genetic innovations, and symbiotic relationships shaped their journey across millions of years.

At the heart of this story lies a series of transformative milestones: the development of a vertebral column, the transition from water to land, and the evolution of complex organ systems such as lungs and limbs. These adaptations not only enabled vertebrates to conquer new environments but also laid the foundation for the emergence of higher cognitive functions, including consciousness. By examining the interplay of cellular biology, environmental challenges, and evolutionary mechanisms, this chapter presents vertebrate evolution as a testament to life's ability to adapt and thrive.

Drawing on insights from cellular biology and evolutionary theory, the chapter begins with the origins of vertebrate traits, tracing their emergence from common ancestors in the primordial seas. It explores the pivotal role of genetic pathways, such as those regulated by parathyroid hormone-related protein (PTHrP), in facilitating physiological adaptations like skeletal stiffening and respiratory efficiency. These molecular mechanisms underscore the dynamic interplay between genes and environments, highlighting the evolutionary significance of homeostatic regulation.

The chapter also considers key transitions, such as the shift from aquatic to terrestrial life, where vertebrates developed innovations like lungs and limbs to overcome the challenges of gravity and desiccation. These transitions underscore the role of pleiotropy, exaptation, and symbiogenesis in evolutionary innovation, illustrating how existing traits can be repurposed to meet new environmental demands.

Finally, the chapter situates vertebrate evolution within a broader philosophical and cosmological context. By linking the adaptive processes of vertebrates to universal principles such as symmetry, entropy, and complexity, it proposes that



J. S. Torday, R. G. Sacco, *Conscious Life*,
https://doi.org/10.1007/978-3-031-80500-4_13

vertebrate evolution mirrors the dynamic forces shaping the cosmos. This perspective not only deepens our understanding of vertebrate biology but also offers a unifying framework for studying the interconnectedness of life and the universe.

The story of vertebrate evolution is not merely a biological narrative; it is a reflection of life's intrinsic creativity and adaptability. As we explore this remarkable journey, we uncover the principles that have guided the development of vertebrates and, by extension, ourselves. Through this lens, vertebrate evolution becomes a profound exploration of the interconnectedness of life, revealing how the forces of nature have shaped the complexity and diversity of the living world.

## Cellular Origins and the Role of Symbiogenesis

The cellular origins of life represent a profound convergence of physics, chemistry, and biology, providing the foundational framework for understanding consciousness. From the unicellular state, life has evolved into complex organisms, driven by processes such as symbiogenesis, which integrates environmental challenges into physiological systems. This section explores how the earliest cellular forms bridged abiotic and biotic realms, highlighting their synchronic (moment-to-moment) and diachronic (evolutionary) dimensions.

### *Consciousness and the Unicellular State*

Consciousness, as a physiological outcome, can be traced back to the unicellular state where life began. The ability of single cells to maintain homeostasis, adapt to environmental changes, and process information forms the basis for what Torday (2023a, b) describes as the synchronic and diachronic dimensions of life. The synchronic dimension captures a cell's real-time ability to sense and respond to its environment, while the diachronic dimension reflects its capacity to encode these adaptations across evolutionary time.

In this context, unicellular organisms serve as the origin of consciousness, embodying the principles of information processing and adaptability. These fundamental traits, which persist in multicellular organisms, highlight the continuity of life from its earliest stages to its most complex forms.

### *Lipids, Micelles, and the Bridge Between Abiotic and Biotic*

The journey from abiotic chemistry to biotic life was facilitated by the unique properties of lipids, molecules that form the structural basis of cellular membranes. Lipids delivered to early Earth via asteroids and comets played a pivotal role in this transition. Deamer (2017) demonstrated how these extraterrestrial lipids spontaneously assembled into micelles under the influence of gravitational and chemical dynamics. Micelles, as self-organizing structures, created semipermeable boundaries capable of encapsulating biochemical reactions, effectively bridging the abiotic and biotic worlds.

This lipid-based innovation provided a stable environment for the emergence of complex molecular interactions, enabling the concentration and organization of molecules essential for life. The ability of micelles to regulate their internal environment marked a significant step toward the evolution of cellular homeostasis, a cornerstone of life's persistence and adaptability.

### *Symbiogenesis: Integrating Environmental Challenges*

Symbiogenesis, the process by which one organism incorporates another into its physiology, serves as a critical mechanism in the evolution of complexity. Sagan (1967) first proposed this concept to explain the origin of eukaryotic cells, where mitochondria and chloroplasts arose from the integration of formerly free-living prokaryotes. This process represents more than a merging of entities; it reflects the assimilation of environmental challenges into adaptive solutions.

Through symbiogenesis, cells developed the ability to internalize external pressures, transforming them into physiological advantages. For instance, mitochondria enabled efficient energy production, allowing eukaryotic cells to thrive in oxygen-rich environments. This integration of external elements into cellular systems exemplifies how life evolves by incorporating and adapting to its surroundings, forming the foundation for cellular memory and evolutionary adaptability.

### *Conclusion*

The origins of life, rooted in the cellular state and shaped by symbiogenesis, underscore the interconnectedness of biology, chemistry, and physics. From the formation of lipid micelles to the integration of environmental challenges through symbiogenesis, these early processes established the physiological and evolutionary basis for consciousness. By linking synchronic cellular functions to diachronic evolutionary trajectories, this framework reveals the profound continuity of life, offering insights into the emergence of complexity and adaptability.

## Evolutionary Milestones: From Aquatic to Terrestrial Life

The transition of vertebrates from aquatic to terrestrial environments represents a defining moment in evolutionary history. This shift required profound physiological adaptations that enabled vertebrates to overcome the challenges posed by life on land, such as desiccation, gravity, and oxygen availability. These adaptations, driven by cellular mechanisms and environmental pressures, exemplify the dynamic interplay between biology and ecology in shaping evolutionary trajectories. This section explores key milestones that facilitated this remarkable transition.

## *Adapting to Terrestrial Life*

The move from water to land required evolutionary innovations to address the challenges of gravity, respiration, and mobility. Among these adaptations, lung alveolarization, skeletal stiffening, and hormonal signaling played pivotal roles. Lungs evolved to efficiently extract oxygen from air, transitioning from the simpler sac-like structures in fish to the highly branched alveolar systems seen in terrestrial vertebrates. This enhanced respiratory efficiency was critical for supporting the higher metabolic demands of life on land (Clack, 2012).

Similarly, skeletal stiffening provided structural support to counteract gravity, enabling vertebrates to maintain upright postures and locomote on land. The evolution of stronger, mineralized bones marked a significant departure from the more flexible skeletal structures of aquatic species. Hormonal signaling, particularly through parathyroid hormone-related protein (PTHrP), played a crucial role in regulating these skeletal changes, ensuring the development of robust and adaptable bone structures (Torday, 2022).

## *Environmental Factors and Physiological Responses*

Environmental factors, particularly oxygen fluctuations, were instrumental in shaping vertebrate physiology during this transition. The evolution of the glucocorticoid receptor, a key regulator of stress responses, exemplifies how vertebrates adapted to variable oxygen availability. This receptor facilitated the regulation of metabolic processes, allowing vertebrates to survive and thrive in fluctuating conditions.

PTHrP further highlights the link between environmental pressures and physiological adaptation. Initially associated with calcium regulation in aquatic environments, PTHrP evolved new functions in terrestrial vertebrates, contributing to bone remodeling and respiratory adaptations. These functional expansions underscore the pleiotropic nature of evolutionary mechanisms, where existing traits are co-opted for new purposes to meet emerging environmental demands (Ben Maamar et al., 2021).

### *The Evolutionary Continuum*

The transition from water to land exemplifies an evolutionary continuum, where cellular and physiological processes respond to environmental dynamics in a seamless progression. Cellular memory and epigenetic inheritance played critical roles in this continuum, enabling organisms to retain adaptive responses across generations. For instance, epigenetic modifications facilitated the integration of environmental information into the genome, ensuring that adaptive traits were passed on to progeny.

This continuum underscores the interconnectedness of life's evolutionary history, where cellular processes serve as the foundation for organismal and ecological adaptations. By linking the molecular mechanisms of lung and skeletal evolution to broader environmental changes, the vertebrate transition to land illustrates the dynamic interplay between biology and environment, highlighting the resilience and adaptability of life.

### *Conclusion*

The evolutionary milestones that enabled vertebrates to transition from aquatic to terrestrial life reflect the intricate relationship between cellular processes and environmental dynamics. From lung alveolarization and skeletal stiffening to hormonal signaling and epigenetic inheritance, these adaptations showcase the ingenuity of evolutionary mechanisms in overcoming challenges. This narrative not only deepens our understanding of vertebrate evolution but also highlights the broader principles of adaptability and interconnectedness that underpin life.

## The Periodic Table and Evolution as Parallel Algorithms

The Periodic Table of Elements, a cornerstone of chemistry, offers profound insights into the mechanisms of evolution. By framing the synchronic (horizontal relationships) and diachronic (temporal evolution) dimensions of the periodic table as homologous to evolutionary processes, parallels emerge between chemical and biological systems. This section explores how the predictive framework of the periodic table mirrors the dynamics of evolutionary biology, highlighting their shared capacity to decode the origins and adaptations of life.

## Synchronic and Diachronic Dimensions

The periodic table's synchronic and diachronic dimensions provide a powerful lens for understanding evolution. In chemistry, synchronic relationships are represented by the horizontal rows (periods), where elements share incremental changes in atomic structure. Diachronic dimensions appear in the periodic trends, reflecting the historical emergence of elements through processes like stellar nucleosynthesis (Scerri, 2007). Similarly, evolutionary biology operates through synchronic mechanisms, such as real-time cellular signaling and adaptation, and diachronic processes like genetic inheritance and phylogenetic divergence (Torday, 2023a, b).

For instance, the diachronic evolution of vertebrates can be compared to the gradual organization of elements within the periodic table. Just as heavier elements emerged through cosmic processes, vertebrate adaptations unfolded in response to environmental pressures, building complexity over time. This analogy underscores the unity of natural laws across disciplines, where both chemical and biological systems are guided by principles of order, balance, and progression.

## Cellular Signaling and Chemical Reactions

Cellular signaling mechanisms bear striking similarities to chemical interactions described in the periodic table. At the cellular level, signaling pathways act as "reaction networks," where molecules interact to produce physiological outcomes. For example,

, essential for processes like muscle contraction and neural communication, parallels the behavior of elements within chemical reactions. These molecular interactions are governed by predictable patterns, akin to Mendeleev's predictive organization of elements (Whitehead, 1929).

Mendeleev's insight into the periodicity of chemical properties resonates with the predictability of evolutionary adaptations. Just as he anticipated the discovery of new elements based on existing patterns, evolutionary biologists can predict potential adaptations by examining genetic and environmental trends. This predictive capacity highlights the shared algorithmic nature of chemical and biological systems, where underlying patterns guide the emergence of complexity.

## Predictive Power and Cosmic Origins

Both the periodic table and evolutionary biology serve as predictive frameworks, offering insights into the origins and adaptations of life. The periodic table's organization is rooted in cosmic events, such as stellar nucleosynthesis, where elements were forged in the cores of stars. These elements later became the building blocks of life, forming the biochemical foundation of cells and organisms. Evolutionary biology, similarly, traces life's adaptations to cosmic influences, from the availability of oxygen to gravitational forces that shaped terrestrial physiology (Torday, 2023a, b).

By linking the chemical origins of elements to biological adaptations, the periodic table and evolutionary biology reveal their interconnectedness. This synthesis not only deepens our understanding of life's origins but also emphasizes the predictive power of these frameworks for exploring future adaptations and evolutionary pathways.

### *Conclusion*

The parallels between the periodic table and evolutionary biology illuminate the shared mechanisms underlying chemical and biological systems. Through their synchronic and diachronic dimensions, both frameworks reveal the dynamic processes that drive complexity and adaptation. By viewing evolution as a parallel algorithm to the organization of elements, this perspective highlights the unity of natural laws, offering predictive insights into the origins and adaptations of life.

## The Interplay of Physics, Chemistry, and Biology in Consciousness

The emergence of consciousness is a complex phenomenon that arises from the interplay of physics, chemistry, and biology. These disciplines converge in cellular processes, with homeostasis acting as a unifying principle. By examining the roles of energy flows, physical forces, and integrative mechanisms, this section explores how consciousness originates from the dynamic integration of local and nonlocal phenomena.

### *Homeostasis: The Unifying Principle*

Homeostasis is the cornerstone of life, representing the dynamic balance that allows cells and organisms to adapt to changing environments. David Bohm (1980) described this process as a manifestation of the implicate and explicate orders, where underlying quantum phenomena unfold into observable realities. In biological systems, homeostasis maintains stability through the regulation of energy flows, molecular signaling, and environmental interactions.

Energy flows are central to this unification, driving cellular and organismal evolution. Margulis (2001) highlighted how symbiotic relationships, such as those leading to the formation of mitochondria, are rooted in energy efficiency and homeostatic regulation. These energy-managing organelles exemplify how biology integrates chemical and physical processes to sustain life. The homeostatic mechanisms that emerged in unicellular organisms laid the foundation for the more complex processes seen in multicellular systems, including the neural networks that underpin consciousness.

## *The Role of Physical Forces in Cellular Processes*

Physical forces, particularly gravity, have a profound impact on cellular processes. Microgravity experiments, such as those conducted by NASA, have demonstrated how the absence of gravitational forces causes cells to lose their phenotypic identity and revert to a more primitive state (Torday, 2003). This reversion underscores the critical role of gravity in shaping cellular organization and differentiation.

Gravity not only influences structural adaptations, such as skeletal stiffening and lung development during the water-to-land transition, but also affects cellular signaling pathways. For example, calcium ion signaling, essential for processes like muscle contraction and neurotransmission, is modulated by gravitational forces. These findings highlight the interdependence of physics and biology, where physical principles like gravity shape the evolution and functionality of biological systems.

## *Consciousness as an Emergent Phenomenon*

Consciousness emerges from the interplay of local and nonlocal phenomena, integrating physical, chemical, and biological processes. Locally, cellular functions like energy metabolism and molecular signaling maintain homeostasis, ensuring survival and adaptability. Nonlocally, quantum phenomena such as entanglement link these cellular processes to broader cosmic dynamics.

This integration suggests that consciousness is not confined to the brain or nervous system but is a holistic property of life. The implicate and explicate orders described by Bohm (1980) provide a framework for understanding this emergence, where quantum-level interactions manifest as complex, adaptive behaviors. By connecting local cellular mechanisms to nonlocal forces, consciousness bridges the microcosm of biology and the macrocosm of the cosmos.

### *Conclusion*

The interplay of physics, chemistry, and biology forms the foundation of consciousness, with homeostasis serving as the central unifying principle. Through energy flows, gravitational influences, and the integration of local and nonlocal phenomena, life evolves toward increasing complexity and awareness. This perspective not only deepens our understanding of consciousness but also underscores the interconnectedness of all natural processes, from the cellular to the cosmic.

## Implications for Understanding Consciousness and Evolution

The synthesis of synchronic and diachronic processes in life forms provides a transformative lens for understanding consciousness and evolution. This section explores the holistic integration of these concepts, linking biology to its cosmic origins, bridging material and energetic perspectives, and proposing a unified framework for comprehending the nature of consciousness.

### *Consciousness as the Integration of Synchronic and Diachronic Processes*

Consciousness can be framed as the ultimate integration of synchronic and diachronic processes. Synchronic processes, such as cellular signaling and homeostasis, represent real-time interactions that sustain life. Diachronic processes, including evolutionary adaptations and genetic inheritance, provide the temporal continuity that enables life to evolve and persist across generations.

Chalmers (1995) described the "hard problem" of consciousness as understanding how subjective experience arises from objective processes. By linking synchronic cellular functions to diachronic evolutionary trajectories, a holistic perspective emerges, situating consciousness as both an emergent and an integrative property of life. Torday (2023a, b) expanded on this by connecting these processes to their cosmic origins, emphasizing that life's adaptive mechanisms reflect universal principles of energy regulation and order.

The interplay between these dimensions suggests that consciousness is not confined to neural systems but is distributed across biological and cosmic scales. This perspective challenges traditional materialist views, positioning consciousness as a fundamental property of life rooted in its evolutionary and physical context.

## *Interdisciplinary Research: Bridging Materialism and Energetic Processes*

Understanding consciousness and evolution requires bridging materialistic and energetic frameworks. Historically, science has focused on material explanations, emphasizing the biochemical and structural aspects of life. While these perspectives provide valuable insights, they often neglect the energetic and informational processes that underpin biological systems.

Whitehead (1929) advocated for a holistic view of reality, proposing that material and energetic phenomena are interconnected. Modern research supports this notion, showing that biological systems operate as dynamic networks where energy flows drive adaptation and evolution. For example, cellular homeostasis relies on energy gradients, such as those generated by mitochondria, to maintain stability and function.

Interdisciplinary research that integrates material and energetic perspectives is essential for advancing our understanding of consciousness and evolution. By combining insights from biology, physics, and philosophy, researchers can develop models that capture the complexity of living systems, bridging gaps between reductionist and holistic approaches.

## *Philosophical and Scientific Challenges*

Comprehending consciousness presents significant philosophical and scientific challenges. Philosophically, the nature of consciousness raises questions about the relationship between mind and matter, the subjective and the objective, and the local and the nonlocal. These dualities require frameworks that can reconcile opposing perspectives, such as Bohm's implicate and explicate orders or Chalmers' dual-aspect theory (1995).

Scientifically, the study of consciousness demands methodologies that account for its multidimensional nature. Traditional approaches that isolate individual variables often fail to capture the emergent properties of complex systems. A unified framework must integrate insights from quantum mechanics, evolutionary biology, and systems theory to provide a comprehensive understanding.

This unified approach has profound implications for fields such as neuroscience, medicine, and cosmology. By framing consciousness as an emergent property of life's synchronic and diachronic processes, researchers can explore its role in adaptation, resilience, and the broader dynamics of the universe.

## *Conclusion*

Consciousness, as the ultimate integration of synchronic and diachronic processes, links biology to its cosmic origins, reflecting the unity of material and energetic dimensions. By bridging disciplinary boundaries and embracing holistic perspectives, researchers can address the philosophical and scientific challenges of understanding consciousness. This approach not only deepens our knowledge of life and evolution but also offers a transformative framework for exploring the fundamental nature of existence.

## References

Ben Maamar, M., Lesage, F., et al. (2021). Epigenetic inheritance and the evolution of vertebrate physiology. *Nature Communications, 12*(1), 4357.

Bohm, D. (1980). *Wholeness and the implicate order*. Routledge.

Chalmers, D. J. (1995). Facing up to the problem of consciousness. *Journal of Consciousness Studies, 2*(3), 200–219.

Clack, J. A. (2012). *Gaining ground: The origin and evolution of Tetrapods*. Indiana University Press.

Deamer, D. (2017). The role of lipid membranes in life's origin. *Life, 7*(1), 5.

Margulis, L. (2001). *Symbiotic planet: A new look at evolution*. Basic Books.

Sagan, L. (1967). On the origin of mitosing cells. *Journal of Theoretical Biology, 14*(3), 255–274.

Scerri, E. R. (2007). *The periodic table: Its story and its significance*. Oxford University Press.

Torday, J. S. (2003). Parathyroid hormone-related protein is a gravisensor in lung and bone cell biology. *Advances in Space Research, 32*, 1569–1576.

Torday, J. S. (2022). Quantum mechanics predicts evolutionary biology. *Progress in Biophysics and Molecular Biology, 135*, 11–15.

Torday, J. S. (2023a). Consciousness, embodied quantum entanglement. *Progress in Biophysics and Molecular Biology, 177*, 125–128.

Torday, J. S. (2023b). Synchronic and diachronic perspectives in the study of consciousness. *Biological Theory, 18*(4), 210–224.

Whitehead, A. N. (1929). *Process and reality*. Macmillan.

# Chapter 14
# Mindfulness in Biology: Cellular Adaptation

The concept of mindfulness, often associated with human consciousness and self-awareness, extends deeply into the biological realm. At the cellular level, mindfulness is embodied in the precise and adaptive responses of cells to their environment. These dynamic processes, honed over billions of years of evolution, illustrate how life maintains equilibrium amidst constant change. Cellular adaptation is the cornerstone of this biological mindfulness, integrating environmental signals, internal homeostasis, and evolutionary pressures to ensure survival and innovation.

This chapter explores the parallels between mindfulness as a cognitive practice and its biological foundations. Just as mindfulness cultivates a heightened awareness of the present moment, cellular processes reflect a moment-to-moment responsiveness to environmental cues, while simultaneously encoding these experiences into a diachronic framework of memory and adaptation. From symbiogenesis as a mechanism for assimilating external challenges into physiological advantages to the cellular basis for healing and systemic resilience, the story of cellular adaptation is one of profound mindfulness at the molecular level.

We will delve into the evolutionary mechanisms that underpin cellular mindfulness, examining the role of feedback loops, epigenetic inheritance, and molecular signaling pathways. Additionally, this chapter will highlight the implications of these processes for understanding health, disease, and human well-being, bridging the gap between cellular biology and the broader philosophical implications of mindfulness.

By drawing on insights from cellular biology, evolutionary theory, and neuroscience, this chapter positions mindfulness not merely as a human cognitive ability but as a fundamental principle of life itself. Cellular adaptation, as a reflection of biological mindfulness, offers profound lessons about resilience, interconnectedness, and the capacity for transformation in the face of adversity.



J. S. Torday, R. G. Sacco, *Conscious Life*,
https://doi.org/10.1007/978-3-031-80500-4_14

## Bridging Mindfulness and Physiology

Mindfulness, a cognitive state often associated with self-awareness and present-moment focus, finds profound parallels within the cellular and molecular mechanisms that govern physiology. At its core, mindfulness reflects the capacity to respond adaptively and dynamically to environmental inputs. This section explores how Ellen Langer's research on mindfulness aligns with the principles of cellular–molecular physiology, illustrating a continuum from individual consciousness to the fundamental processes of life itself.

### *Energy Flows: The Foundation of Mindfulness*

Both mindfulness and cellular physiology are underpinned by the concept of energy flow. In mindfulness, energy refers to the allocation of cognitive and attentional resources to perceive, process, and respond to the present moment. Similarly, cellular processes operate through the flow of biochemical energy, such as ATP production and utilization, to maintain homeostasis. This energy-driven adaptability connects classical physics, which governs macroscopic processes, with quantum mechanics, where probabilistic phenomena guide molecular interactions (Torday & Miller, 2018).

At the cellular level, energy flows mediate critical functions such as signaling, repair, and growth, mirroring the adaptive qualities of mindfulness. This perspective suggests that mindfulness, rather than being a solely human cognitive capacity, is a universal principle of life, rooted in the energy flows that sustain biological systems. Cellular homeostasis, for instance, reflects an intrinsic "mindfulness" that allows cells to adapt dynamically to fluctuating environments while preserving internal stability.

### *Symbiogenesis: Linking Cells to Organismal Mindfulness*

Symbiogenesis provides a compelling framework for understanding how cellular mechanisms scale up to organismal mindfulness. Initially described by Lynn Margulis (Sagan, 1967), symbiogenesis posits that the integration of once-independent entities, such as mitochondria and host cells, formed the basis for complex life. This process not only established the foundation for cellular energy efficiency but also created feedback loops that coordinate larger physiological systems.

These symbiotic relationships exemplify the interconnectedness inherent in mindfulness. At the cellular level, communication networks between organelles, such as mitochondria and the nucleus, ensure that cells respond holistically to external stimuli. These same principles extend to tissues and organ systems, which integrate local cellular processes into cohesive organismal responses. The result is a continuum of mindfulness, where cells, tissues, and organisms operate adaptively to maintain homeostasis in the face of environmental challenges (Torday, 2023).

### *Ellen Langer's Mindfulness Studies: Insights into Physiology*

Ellen Langer's pioneering research provides empirical evidence for the physiological effects of mindfulness. Her studies demonstrate that mindset—the cognitive framing of experiences—can significantly influence physiological states. For example, her iconic study with hotel maids revealed that simply reframing their daily physical activities as exercise improved their health metrics, including weight, body fat, and blood pressure (Langer, 2007). This finding underscores the bidirectional relationship between cognitive awareness and physiological function, mirroring the adaptive interplay seen in cellular processes.

Similarly, Langer's octogenarian study, in which elderly participants were immersed in an environment reminiscent of their youth, showed remarkable reversals in age-related physical and cognitive decline (Pagnini et al., 2019). These studies highlight how mindfulness impacts energy flow and homeostasis, further bridging the cognitive and physiological dimensions of adaptation.

### *Conclusion*

The convergence of mindfulness and cellular–molecular physiology reveals a deep interconnectedness between cognitive states and biological processes. By grounding mindfulness in energy flows and highlighting the role of symbiogenesis, this perspective positions mindfulness as a universal principle that transcends the boundaries of individual consciousness. Ellen Langer's research exemplifies the profound impact of mindset on physiological adaptation, offering insights into how mindfulness operates at every scale of life, from cells to organisms.

## Cellular Communication and Evolutionary Adaptation

The ability to adapt to environmental changes and maintain internal stability is central to life. At its core, this adaptability is underpinned by cell–cell communication, which evolved as a foundational mechanism for homeostasis and survival. This section explores how cellular communication mirrors the principles of mindfulness, emphasizing its role in evolutionary adaptation, symbiogenesis, and the aging process.

## *Cell–Cell Communication: The Foundation of Physiological Mindfulness*

Cell–cell communication evolved as a primary mechanism for maintaining homeostasis, the dynamic balance of internal conditions essential for life. This process involves intricate signaling pathways that enable cells to perceive and respond to environmental changes. By coordinating their activities, cells ensure that tissues and organs function cohesively, mirroring the real-time awareness characteristic of mindfulness.

Torday and Rehan (2012) highlighted how this communication is fundamental to all multicellular organisms. For example, gap junctions allow direct cytoplasmic exchanges between adjacent cells, facilitating the synchronization of metabolic and signaling activities. Similarly, paracrine signaling ensures that cells communicate over short distances to regulate processes such as inflammation and repair. These cellular interactions reflect a form of physiological mindfulness, where the "awareness" of individual cells contributes to the organism's overall adaptability and resilience.

## *Symbiogenesis: Integrating Threats into Adaptive Systems*

Symbiogenesis, the evolutionary process by which separate organisms form symbiotic relationships, represents a profound example of cellular adaptation. Initially proposed by Lynn Margulis, this process illustrates how cells have historically assimilated environmental threats into their physiology to enhance survival (Case et al., 2007). For instance, the integration of mitochondria into ancestral eukaryotic cells provided a significant evolutionary advantage, enabling efficient energy production and robust homeostasis.

This process parallels mindful adaptation, where external challenges are internalized and transformed into strengths. The incorporation of mitochondria not only optimized cellular energy flow but also established feedback loops that continue to shape physiological processes today. These symbiotic relationships exemplify the mindful integration of environmental information into cellular systems, creating a continuum of adaptation that spans molecular, cellular, and organismal levels.

## *Mindfulness and the Aging Process*

The decline in cellular communication is a hallmark of aging, leading to diminished energy flow and impaired homeostasis. Understanding aging through the lens of cell communication reveals how mindfulness practices can mitigate senescence by enhancing cellular adaptability and energy management.

Storr et al. (2013) described how oxidative stress disrupts cellular signaling pathways, contributing to aging and chronic disease. Mindfulness practices, such as meditation and stress reduction, have been shown to counteract these effects by reducing systemic inflammation and improving mitochondrial function. Kotter-Grühn et al. (2009) further emphasized the role of mindset in influencing physiological aging, highlighting how positive mental states can slow senescence by fostering cellular resilience.

By enhancing cellular communication and energy flow, mindfulness practices offer a practical approach to addressing the physiological underpinnings of aging. This connection underscores the deep interplay between cognitive states and biological processes, bridging the gap between mindfulness and cellular adaptation.

### *Conclusion*

Cellular communication serves as the foundation for evolutionary adaptation and physiological mindfulness. Through mechanisms like gap junctions and symbiogenesis, cells have developed sophisticated systems to maintain homeostasis and respond to environmental challenges. Understanding aging as a decline in these processes highlights the potential of mindfulness practices to enhance cellular resilience and slow senescence. This perspective reveals mindfulness not only as a cognitive practice but also as a biological principle embedded in the very fabric of life.

## The Role of Environment in Mindful Adaptation

Environmental factors profoundly shape the adaptive capacities of both cells and organisms, guiding their responses to external challenges. This section explores the dynamic interplay between environment and biology, emphasizing how adaptation mirrors mindfulness through processes such as niche construction, gravity-driven, and epigenetic regulation. These mechanisms reveal the profound influence of the environment on physiological and behavioral resilience, paralleling the mindfulness practices studied by Ellen Langer.

### *Niche Construction: Symbiotic Adaptation and Mindfulness*

Niche construction describes the process by which organisms actively modify their environments to improve their survival, creating a feedback loop between adaptation and environmental change. This concept aligns with Langer's mindfulness research, which emphasizes the active role of perception and mindset in shaping reality. Just as organisms transform their environments to reduce stress and enhance survival, individuals can alter their mental landscapes to foster resilience and well-being (Torday, 2016; Sagan, 1967).

At the cellular level, niche construction occurs through symbiotic adaptation. For example, mitochondria, originally independent prokaryotes, were incorporated into eukaryotic cells, forming an energy-efficient system that adapted to environmental challenges like oxygen fluctuations. This process exemplifies how biology mirrors mindful adaptation by integrating external stressors into functional advantages. Langer's studies on workplace mindfulness further support this concept by demonstrating that shifts in perception can significantly impact physical and mental health, akin to how symbiotic relationships optimize cellular efficiency.

## *Gravity as a Driver of Cellular Evolution*

Gravity, one of the fundamental forces of nature, has played a critical role in shaping the evolution of life. From the alignment of lipid micelles in early protocells to the development of complex skeletal structures in terrestrial organisms, gravity has influenced cellular and organismal adaptation. These gravitational effects reflect a form of mindful physiological adaptation, where life responds to environmental forces to maintain equilibrium (Torday, 2003).

For instance, microgravity experiments have shown that cells lose their phenotypic identity in the absence of gravitational cues, reverting to a more primitive state. This finding underscores the importance of environmental inputs in maintaining cellular structure and function. Mindful adaptation at the organismal level reflects a similar principle, where physical and psychological states are shaped by environmental pressures. Langer's mindfulness interventions, which focus on altering environmental perceptions, echo this adaptive process by promoting health and resilience.

## *Epigenetics: Mediators of Environmental Influence*

Epigenetic mechanisms provide a cellular explanation for how environmental factors influence physiology and behavior. By modifying gene expression without altering the DNA sequence, epigenetics enables organisms to adapt to environmental changes dynamically. This flexibility is central to the placebo effect, where beliefs and expectations can induce measurable physiological changes (Colloca & Barsky, 2020).

Ellen Langer's experiments demonstrate the behavioral and physiological shifts that occur when individuals adopt a mindful mindset. For example, her study on hotel maids revealed that reframing their daily activities as exercise led to improved health metrics, illustrating the power of perception in mediating biological responses. Epigenetic mechanisms provide a cellular basis for these effects, linking environmental cues to changes in gene expression and systemic adaptation (Ben Maamar et al., 2021).

### *Conclusion*

The role of the environment in mindful adaptation underscores the interconnectedness of biology, behavior, and external conditions. Through niche construction, gravity-driven evolution, and epigenetic regulation, life demonstrates an intrinsic capacity for dynamic adaptation. These processes align with the principles of mindfulness, showing how perception and environmental interaction shape both cellular and organismal resilience. Langer's studies highlight the transformative potential of mindful practices, revealing their deep biological roots and their capacity to foster well-being in an ever-changing world.

## Mindfulness as Cellular Homeostasis

Mindfulness, often understood as a state of present-moment awareness, finds a profound biological parallel in the concept of homeostasis. Cellular homeostasis, the process by which cells maintain internal stability in response to external stimuli, mirrors the mindful ability to adapt dynamically to environmental changes while preserving equilibrium. This section frames mindfulness as an extension of cellular homeostasis, exploring its roots in physiological processes and its implications for health and well-being.

### *Homeostasis as the Foundation of Mindfulness*

The principle of homeostasis, first described by Claude Bernard and later elaborated by Walter Cannon, refers to the biological mechanisms that regulate the internal environment of an organism. Cellular homeostasis involves maintaining stable conditions, such as pH, ion concentrations, and energy levels, despite external fluctuations. This adaptability reflects a form of "awareness" at the cellular level, where biochemical processes respond precisely to environmental inputs to sustain life (Mitchell, 1961; Torday, 2015).

This concept parallels mindfulness, where an individual's conscious awareness allows for adaptive responses to external stimuli. At the cellular level, homeostasis operates through feedback mechanisms that restore balance, akin to how mindfulness practices recalibrate psychological and physiological states. For example, cells repair oxidative damage and maintain redox balance through mechanisms like the activation of antioxidant pathways, demonstrating a micro-level mindfulness that supports survival (Storr et al., 2013).

## Mindset and Cellular Adaptation

Ellen Langer's hypothesis that "mindset determines health" aligns with the cellular principles of adaptation and homeostasis. Langer's studies have shown that shifts in perception can lead to measurable physiological changes, suggesting that the mind plays a critical role in influencing bodily states. At the cellular level, this idea finds support in the ability of cells to adapt dynamically to stressors, such as oxidative stress, by activating repair mechanisms.

Oxidative stress occurs when reactive oxygen species (ROS) overwhelm the cell's antioxidant defenses, leading to potential damage to DNA, proteins, and lipids. Cells counteract this imbalance by upregulating antioxidants like glutathione and catalase, restoring homeostasis. Similarly, mindfulness practices have been shown to reduce systemic inflammation and oxidative stress, promoting resilience at both cellular and systemic levels (Torday, 2016; Storr et al., 2013). These processes illustrate how mindful awareness mirrors the cellular capacity for balance and adaptation.

## The Role of Breathing in Homeostasis and Mindfulness

Breathing is a physiological bridge between cellular homeostasis and mindful practice. Controlled breathing techniques, such as diaphragmatic breathing and pranayama, have been shown to regulate heart rate, blood pressure, and stress responses, contributing to systemic balance. At the cellular level, breathing modulates oxygen delivery, ensuring that cells receive the necessary resources to maintain homeostasis during stress (Balban et al., 2023; Litman, 2023).

For example, slow, deep breathing stimulates the vagus nerve, activating the parasympathetic nervous system and promoting relaxation. This physiological state facilitates homeostasis by reducing the production of stress hormones like cortisol and enhancing antioxidant defenses. These effects underscore the role of breathing as a tool for achieving mindfulness, linking conscious practice to cellular mechanisms.

## Conclusion

Mindfulness and homeostasis share a foundational principle: the ability to maintain balance and stability in the face of change. Cellular homeostasis provides a biological framework for understanding how mindfulness practices promote health and resilience. By connecting mindset, oxidative stress repair, and breathing to cellular adaptation, this perspective highlights the profound interplay between conscious awareness and physiological regulation, revealing mindfulness as a reflection of life's intrinsic capacity for balance.

## Evolutionary Perspectives on Mindfulness

Mindfulness, often associated with present-moment awareness, can also be viewed as an evolutionary and physiological adaptation rooted in survival mechanisms. This perspective positions mindfulness as an emergent property of evolutionary processes, influencing both individual well-being and broader ecological systems. By integrating evolutionary biology, Gaia theory, and modern studies on health and aging, this section explores the transformative potential of mindfulness from a biological and ecological perspective.

### *Mindfulness as an Evolutionary Survival Mechanism*

Evolutionary biology offers a compelling framework for understanding mindfulness as a survival strategy. In its essence, mindfulness involves the ability to adapt dynamically to external changes while maintaining internal stability—a principle deeply embedded in life's evolutionary trajectory. Early life forms, such as unicellular organisms, employed mechanisms akin to mindfulness to respond to environmental stimuli, ensuring homeostasis and survival (Torday, 2016).

Mindfulness also plays a critical role in niche construction, where organisms actively shape their environments to reduce stressors and enhance survival opportunities. For instance, early life forms modified their environments by producing oxygen as a byproduct of photosynthesis, creating conditions conducive to their proliferation. Similarly, mindfulness practices in humans, such as focused attention and stress management, mirror this adaptive capacity by fostering resilience and reducing the physiological impacts of stress (Sagan, 1967; Torday, 2016).

### *Gaia Theory and Mindfulness at the Ecosystem Level*

Gaia theory, proposed by James Lovelock, presents Earth as a self-regulating system where living organisms and their environment interact to maintain planetary homeostasis. This concept aligns with the principles of mindfulness, which emphasize interconnectedness and balance. Just as the Gaia system adapts to maintain equilibrium, mindfulness fosters harmony within individuals and their surroundings, promoting both personal and ecological well-being (Lovelock, 2000).

The holistic effects of mindfulness extend beyond individuals to influence ecosystems. For example, mindful agricultural practices, such as permaculture, mimic natural systems and emphasize sustainability. These practices reflect the principles of Gaia theory, demonstrating how mindfulness-inspired approaches can address ecological challenges. By viewing mindfulness as an extension of Earth's self-regulating processes, we gain insights into its potential to enhance the health of both individuals and the planet (Torday & Rehan, 2012).

## *Implications for Health, Aging, and Well-Being*

Mindfulness, as a physiological and evolutionary process, has profound implications for health, aging, and well-being. Studies have shown that mindfulness practices can slow biological aging by reducing stress and inflammation, enhancing cellular repair mechanisms, and improving overall resilience. This aligns with evolutionary principles, where adaptability and efficient resource use are key to longevity (Langer, 2023).

Furthermore, mindfulness has been linked to epigenetic inheritance, where environmental influences shape gene expression across generations. For example, stress reduction through mindfulness can lead to epigenetic modifications that improve health outcomes for both individuals and their offspring. These findings highlight the evolutionary significance of mindfulness as a mechanism for promoting health and well-being across the lifespan (Torday, 2023).

## *Conclusion*

Viewing mindfulness through an evolutionary lens reveals its role as a fundamental survival mechanism, deeply embedded in the fabric of life. From niche construction to Gaia theory, mindfulness reflects the interconnectedness of biological and ecological systems, offering insights into health, aging, and sustainability. By integrating mindfulness into daily practices and ecological strategies, we can harness its evolutionary potential to enhance both personal and planetary well-being.

## References

Balban, M. Y., Porter, C. A., Lin, J., & Vago, D. R. (2023). The physiological basis of mindful breathing: Impacts on emotional regulation and health. *Mindfulness and Health Journal, 12*(3), 142–156.

Ben Maamar, M., Sadler-Riggleman, I., Beck, D., McBirney, M., Nilsson, E., Klukovich, R., & Skinner, M. K. (2021). Epigenetic transgenerational inheritance of altered sperm histone retention sites. *Scientific Reports, 11*, 3583.

Case, R. M., Eisner, D., Gurney, A., Jones, O., Muallem, S., & Verkhratsky, A. (2007). Evolution of calcium homeostasis: From birth of the first cell to an omnipresent signaling system. *Cell Calcium, 42*(4–5), 345–350.

Colloca, L., & Barsky, A. J. (2020). Placebo and nocebo effects. *The New England Journal of Medicine, 382*(6), 554–561.

Kotter-Grühn, D., Grühn, D., & Smith, J. (2009). Predicting one's own death: The relationship between subjective and objective nearness to death in very old age. *European Journal of Aging, 6*(1), 13–23.

Langer, E. J. (2007). *Counterclockwise: Mindful health and the power of possibility*. Ballantine Books.

Langer, E. (2023). *Mindfulness in action: Bridging the gap between awareness and well-being*. Cambridge University Press.

Litman, L. (2023). Breathing for balance: How controlled breathing techniques enhance resilience. *Journal of Integrative Medicine, 15*(4), 211–223.

Lovelock, J. (2000). *The ages of Gaia: A biography of our living earth*. Oxford University Press.

Mitchell, P. (1961). Coupling of phosphorylation to electron and hydrogen transfer by a chemiosmotic type of mechanism. *Nature, 191*, 144–148.

Pagnini, F., Bercovitz, K. E., & Langer, E. J. (2019). Perceived control and mindfulness: Implications for clinical practice. *Frontiers in Psychology, 10*, 281.

Sagan, L. (1967). On the origin of mitosing cells. *Journal of Theoretical Biology, 14*(3), 255–274.

Storr, S. J., Woolston, C. M., Zhang, Y., & Martin, S. G. (2013). Redox environment, free radical, and oxidative DNA damage. *Antioxidants & Redox Signaling, 18*(18), 2399–2408.

Torday, J. S. (2003). Parathyroid hormone-related protein is a gravisensor in lung and bone cell biology. *Advances in Space Research, 32*, 1569–1576.

Torday, J. S. (2015). Homeostasis as the mechanism of evolution. *Biology, 4*, 573–590.

Torday, J. S. (2016). The cell as the first niche construction. *Biology, 5*, 19.

Torday, J. S. (2023). Ontogeny, phylogeny and cellular energy flows for evolution. *Progress in Biophysics and Molecular Biology, 178*, 78–82.

Torday, J. S., & Miller, W. B. (2018). A systems approach to physiologic evolution. *Journal of Cellular Physiology, 233*, 162–167.

Torday, J. S., & Rehan, V. K. (2012). *Evolutionary biology, cell-cell communication, and complex disease*. Wiley.

# Chapter 15
# Holistic Evolution: The Unity of Life and Consciousness

The story of life is one of unification, a seamless continuum where the biological, quantum, and cosmic intersect to shape the essence of existence. In this chapter, we explore the profound interconnectedness of evolution and consciousness, framing them as two sides of the same coin—one representing the structural and functional progression of life, and the other embodying its emergent, self-aware dimension. Through the lens of holistic evolution, we uncover how the unity of life and consciousness is not merely a philosophical abstraction but a biological reality rooted in the mechanisms of adaptation, cellular communication, and environmental interplay.

From the humble origins of unicellular life to the intricate complexity of multicellular organisms, evolution demonstrates a recurring theme: the ability of life to integrate and adapt in response to environmental challenges. This chapter examines how processes like symbiogenesis, genetic memory, and homeostasis underscore the unity of living systems, fostering a continuous dialogue between the internal and external worlds. The phenomenon of consciousness, often seen as distinct from biological processes, is reframed as an emergent property of this evolutionary dance—a byproduct of the same integrative mechanisms that sustain life.

We begin by delving into the cellular origins of holism, where single cells acted as microcosms of life's potential, bridging the explicate and implicate orders of existence. From there, we trace the evolutionary transitions that propelled organisms toward greater complexity, culminating in the emergence of neural systems capable of higher-order consciousness. Along the way, we examine the role of niche construction, epigenetics, and Gaia theory in shaping both life and consciousness as interconnected phenomena.

Holistic evolution also invites us to consider the unity of life and consciousness within a broader cosmic framework. Cellular processes such as homeostasis and energy flow mirror universal principles like entropy regulation and quantum coherence. These parallels suggest that consciousness is not an isolated phenomenon but

 
J. S. Torday, R. G. Sacco, *Conscious Life*,
https://doi.org/10.1007/978-3-031-80500-4_15

part of a larger pattern that integrates biological systems with the fundamental laws of physics and cosmology.

Finally, this chapter reflects on the implications of holistic evolution for understanding the nature of existence. By embracing the unity of life and consciousness, we can approach pressing questions in science, medicine, and philosophy with a more integrative mindset. From advancing preventive medicine to deepening our understanding of the mind–body connection, holistic evolution offers a pathway to bridging disciplines and exploring the profound interconnectedness that defines our universe.

As we journey through the themes of this chapter, we invite readers to reconsider the boundaries between biology and philosophy, structure and meaning, life and consciousness. By viewing evolution as a holistic process, we can begin to grasp the intricate web of relationships that binds all forms of existence into a unified whole.

## The Evolutionary Leap to Consciousness

The evolutionary leap to human consciousness, often referred to as the "great leap forward," represents a critical juncture in the history of life on Earth. This transformative period, marked by advancements in cognition, bipedalism, and tool use, underscores the intricate relationship between physiology and behavior. By examining key physiological changes, such as oxygen regulation, skeletal adaptations, and neural development, we can better understand how human consciousness emerged as a complex, adaptive phenomenon.

### *The "Great Leap Forward" and Cognitive Evolution*

The "great leap forward," a term coined by Jared Diamond, describes the period when humans exhibited a dramatic increase in cognitive capabilities. This leap is closely tied to genetic changes, particularly those associated with the FOXP2 gene, which plays a critical role in language development and neural plasticity (Diamond, 1999; Xu et al., 2018). Mutations in FOXP2 are thought to have facilitated the emergence of complex language systems, enabling humans to communicate abstract ideas and cooperate at unprecedented levels.

These advancements in cognition were paralleled by the development of sophisticated tool use, another hallmark of early human evolution. The coordination and planning required for toolmaking reflect the expansion of neural circuits within the prefrontal cortex, suggesting a direct link between physiology and behavioral innovation. The interplay of genetic and neural factors underscores how evolutionary pressures shaped the human brain, driving the development of consciousness as a tool for survival.

## From Aquatic to Terrestrial Life: Oxygen and Skeletal Adaptations

The evolutionary transition from aquatic to terrestrial life laid the groundwork for many of the physiological changes that enabled human consciousness. During this transition, oxygen availability and fluctuations played a significant role in shaping vertebrate physiology. For instance, the development of lungs and the process of skeletal calcification were critical for survival on land, where oxygen was more readily available but required efficient respiratory systems to be utilized effectively (Clack, 2012; Torday, 2015).

Skeletal calcification not only provided structural support for terrestrial movement but also created a reservoir for calcium, a key signaling molecule in cellular communication. These adaptations highlight how environmental pressures influenced the evolution of homeostatic mechanisms that supported increasingly complex behaviors. Oxygen fluctuations during this period also drove the evolution of metabolic pathways, providing the energy necessary for advanced neural functions.

## The Role of Bipedalism in Cognitive and Neural Development

The adoption of bipedalism represents one of the most significant evolutionary shifts, freeing the forelimbs for activities such as tool use and gestural communication. Bipedalism also restructured the human skeletal system, aligning the spinal column with the skull and creating a more energy-efficient mode of locomotion (Rodman & McHenry, 1980). This energy efficiency allowed for increased allocation of metabolic resources to the central nervous system, contributing to the enlargement and specialization of the brain (Torday, 2024).

The liberated forelimbs, in turn, facilitated the fine motor skills required for toolmaking and the gestural precursors to language. These behavioral changes likely coevolved with neural adaptations, forming a feedback loop that accelerated cognitive and cultural evolution. The interplay between bipedalism, neural development, and social behavior underscores the role of physiology in shaping human consciousness.

## Conclusion

The evolutionary leap to human consciousness is a story of adaptation and integration. From the genetic innovations associated with the FOXP2 gene to the physiological changes driven by oxygen availability and bipedalism, this transformative period highlights the intricate connections between biology and behavior. By examining these evolutionary milestones, we gain deeper insights into the processes that gave rise to human consciousness, providing a foundation for understanding its complexities in both scientific and philosophical contexts.

## Holism in Evolutionary Adaptation

Holism, the principle that systems and their properties should be viewed as wholes rather than merely as collections of parts, offers a powerful lens for understanding evolutionary adaptation. Contrasting sharply with reductionist views that emphasize random mutations as the primary drivers of evolutionary change, holism reveals the symbiotic and integrative processes that guide the evolution of life. By examining evolutionary mechanisms such as symmorphosis, physiological optimization, and responses to environmental stress, this section underscores how holistic principles shape the adaptive trajectory of organisms.

### *Evolution as a Holistic, Symbiotic Process*

The traditional view of evolution, which frames random mutations and natural selection as the primary drivers of change, fails to capture the integrative and symbiotic nature of adaptation. Lynn Margulis's theory of symbiogenesis challenges this reductionist perspective, demonstrating that cooperative interactions between organisms and their environments play a fundamental role in evolutionary innovation (Sagan, 1967). These interactions enable organisms to integrate environmental challenges into their physiology, creating systems that are optimized for survival.

Stephen Jay Gould and Elisabeth Vrba expanded this view with their concept of exaptation, which highlights how existing traits can be co-opted for new functions in response to changing environments (Gould & Vrba, 1982). For instance, feathers, which originally evolved for thermoregulation, were later adapted for flight in birds. These examples illustrate how evolution operates as a dynamic and holistic process, with traits and systems emerging from a complex interplay of environmental pressures and biological responses.

### *Symmorphosis: Optimizing Physiological Systems*

The concept of symmorphosis, introduced by Weibel and colleagues, provides a framework for understanding how physiological systems are "engineered" to meet environmental demands. Symmorphosis posits that the design of biological systems reflects a balance between structural complexity and functional efficiency, minimizing waste while maximizing performance (Weibel et al., 1998).

The vertebrate lung is a prime example of this principle. As terrestrial life evolved, the lung underwent adaptations to optimize oxygen exchange in increasingly variable environments. The alveolar surface area and capillary network were refined to support efficient gas exchange, reflecting the principle of symmorphosis in action. These structural changes highlight how evolutionary adaptation is not a random process but a holistic response to environmental challenges.

### Adaptive Mechanisms and Hypoxia-Induced Stress

Environmental stressors such as hypoxia (low oxygen availability) play a significant role in driving physiological adaptation. Hypoxia induces a cascade of cellular and systemic responses, promoting traits that enhance metabolic efficiency and respiratory function. For example, organisms exposed to high-altitude environments develop increased red blood cell counts, larger lung volumes, and enhanced oxygen transport capacity, enabling them to thrive under oxygen-limited conditions (Berner et al., 2007).

At the cellular level, hypoxia triggers the activation of hypoxia-inducible factors (HIFs), which regulate genes involved in energy metabolism, angiogenesis, and cellular survival (Rubin et al., 2004). These responses exemplify the holistic nature of adaptation, as organisms integrate environmental cues into their physiological processes to maintain homeostasis and ensure survival.

### Conclusion

Holism provides a comprehensive framework for understanding evolutionary adaptation, emphasizing the integrative processes that drive the optimization of biological systems. From the symbiotic origins of life to the principles of symmorphosis and the adaptive responses to hypoxia, evolution emerges as a dynamic and interconnected process. By adopting a holistic perspective, we gain deeper insights into the mechanisms that shape life's complexity and resilience, offering a richer understanding of evolution's guiding principles.

## Symbiogenesis and Cellular Communication

Symbiogenesis, the integration of formerly independent organisms into a new, cooperative entity, serves as a cornerstone of evolutionary innovation. It provides a mechanism for assimilating environmental stressors into physiology while conserving evolutionary history through cellular signaling networks. This section explores how symbiogenesis and cellular communication have driven major evolutionary milestones, including the transition from aquatic to terrestrial life, and emphasizes the role of epigenetic inheritance in shaping adaptive change across generations.

### Symbiogenesis: Integrating Environmental Stressors

First proposed by Lynn Margulis, symbiogenesis describes the evolutionary process by which symbiotic relationships between distinct organisms lead to the emergence of new biological systems (Sagan, 1967). This process allows organisms to integrate environmental challenges into their physiology, creating robust systems capable of responding to dynamic environmental conditions. Symbiogenesis not only drives evolutionary adaptation but also preserves evolutionary history by embedding the lessons of past challenges into cellular communication networks (Gray, 2017).

For example, mitochondria, essential for cellular respiration, originated from symbiotic relationships between ancestral eukaryotes and free-living bacteria. This integration facilitated efficient energy production, enabling the development of multicellular organisms. Similarly, the endosymbiotic relationships that led to chloroplasts in plants demonstrate how symbiogenesis provides the physiological tools necessary for adapting to new environmental niches.

## *PTHrP: A Key Mediator in the Transition to Land*

The transition from water to land represents one of the most significant evolutionary shifts, requiring organisms to adapt to entirely new environmental stressors. Parathyroid hormone-related protein (PTHrP) has been identified as a key mediator in this process, playing a pivotal role in skeletal calcification and lung alveolarization (Torday & Rehan, 2003).

PTHrP facilitates the development of the skeletal structure required to support terrestrial locomotion, ensuring that organisms can withstand the gravitational stresses of life on land. At the same time, PTHrP regulates the formation of alveoli in the lungs, optimizing oxygen exchange in an air-based environment (Clemens et al., 2001). These dual functions highlight the integrative nature of symbiogenesis and cellular signaling, demonstrating how evolutionary adaptations are deeply interconnected across physiological systems.

## *Epigenetic Inheritance and Adaptive Change*

Epigenetic inheritance extends the influence of environmental factors beyond individual organisms, allowing adaptive changes to be passed across generations. This phenomenon ensures that phenotypes, shaped by environmental conditions, serve as agents of evolutionary change (Torday, 2015).

Epigenetic mechanisms, such as DNA methylation and histone modification, enable organisms to respond rapidly to environmental stressors without requiring genetic mutations. For instance, studies have shown that organisms exposed to environmental toxins can pass on adaptive epigenetic changes to their offspring, enhancing their resilience to similar challenges (Ben Maamar et al., 2021). This ability to integrate environmental data into heritable traits exemplifies the synergy between cellular communication and evolutionary adaptability.

### *Conclusion*

Symbiogenesis and cellular communication are essential drivers of evolutionary innovation, enabling organisms to integrate environmental stressors into their physiology while preserving evolutionary history. From the role of PTHrP in facilitating the transition to land to the broader implications of epigenetic inheritance, these processes demonstrate the interconnectedness of life and the mechanisms that sustain its adaptability. By understanding these fundamental principles, we can better appreciate the intricate dance between organisms and their environments that shapes the trajectory of evolution.

## Breathing and the Physiology of Consciousness

Breathing, a fundamental physiological process, plays a critical role in maintaining homeostasis and influencing consciousness. Its centrality to both survival and cognitive evolution suggests deep evolutionary roots that link respiration to higher-order functions. This section explores the intricate relationship between breathing and consciousness, examining its biochemical, evolutionary, and philosophical dimensions.

### *Breathing and Biochemical Pathways*

Controlled breathing practices have profound effects on biochemical pathways, altering states of consciousness. These practices, such as deep diaphragmatic breathing, have been shown to modulate stress hormones like cortisol, reducing stress and enhancing mental clarity. For example, studies have demonstrated that controlled breathing can inhibit cortisol production, fostering states of calm and focused awareness (Bing-Canar et al., 2016).

Furthermore, controlled breathing may stimulate the endogenous production of dimethyltryptamine (DMT), a compound associated with altered states of consciousness and mystical experiences. Stanislav Grof (2010) proposed that breathing techniques, such as holotropic breathing, can facilitate access to expanded states of awareness by activating latent neurochemical pathways. These findings link respiration to the regulation of neurochemical states, positioning it as a bridge between physiological function and consciousness.

## Breathing and Evolutionary Adaptation

From an evolutionary perspective, breathing is deeply intertwined with the development of bipedalism and higher cognition. The transition to bipedal locomotion required significant changes in respiratory mechanics, as the thoracic cavity evolved to accommodate both the physical demands of upright posture and the metabolic needs of an enlarged brain (Torday, 2015).

This evolutionary adaptation underscores the dual role of breathing in supporting both physical and cognitive development. Deep, rhythmic breathing not only enhances oxygen delivery to tissues but also regulates the autonomic nervous system, maintaining homeostasis under variable environmental conditions. These physiological mechanisms highlight the evolutionary importance of breathing as a tool for survival and adaptive flexibility.

## Breathing in Eastern Philosophy and Modern Psychology

Breathing practices have long been recognized in Eastern philosophical traditions, such as yoga and pranayama, as pathways to heightened states of consciousness. These practices emphasize the integration of body and mind through controlled respiration, aligning physiological processes with mental states. Maturana and Varela (1972) framed such integrative practices as reflective of the unity of the living system, where respiration embodies a continuous exchange between the internal and external environment.

Modern psychology has also embraced the role of breathing in mental health. Techniques such as mindfulness-based stress reduction (MBSR) incorporate controlled breathing to improve emotional regulation and resilience. These practices resonate with evolutionary physiology, suggesting that the roots of breathing's impact on consciousness lie in its ability to stabilize internal states in response to external stressors (Torday, 2023).

## Conclusion

Breathing represents a convergence of evolutionary, biochemical, and philosophical dimensions, serving as a vital mechanism for maintaining homeostasis and influencing consciousness. By exploring its physiological and cultural significance, we gain deeper insights into the role of respiration as a cornerstone of both survival and self-awareness. This synthesis of ancient practices and modern science underscores breathing's enduring relevance in understanding the physiology of consciousness.

## Holism, Homeostasis, and Consciousness

Holism and homeostasis are foundational concepts for understanding the evolution of consciousness. Consciousness, as an emergent property of life, is deeply rooted in the evolutionary processes that prioritize the regulation of energy and the maintenance of internal balance. This section synthesizes these principles to address the evolution of consciousness, its philosophical implications, and its applications in contemporary challenges.

### *Consciousness as an Emergent Property*

Consciousness arises from the interplay of evolutionary mechanisms that emphasize homeostasis and energy regulation. Maturana and Varela (1972) proposed that the self-organizing nature of living systems inherently integrates physiological processes with cognitive functions. Cellular homeostasis, the ability to maintain stable internal conditions amidst external fluctuations, provides the biological foundation for consciousness.

From an evolutionary perspective, homeostasis serves as a survival mechanism that ensures the continuity of life. The transition from unicellular to multicellular organisms involved increasingly complex systems for maintaining internal stability. These systems include the evolution of the endocrine network, which integrates hormonal signals with metabolic processes to support energy flow and adaptation (Torday, 2019). This integration frames consciousness as a byproduct of evolutionary strategies for managing energy and environmental challenges.

### *Holism and the "Hard Problem" of Consciousness*

David Chalmers (1995) defined the "hard problem" of consciousness as the challenge of explaining how subjective experiences (qualia) arise from physical processes. Holism offers a framework for addressing this issue by viewing consciousness as the product of interconnected systems rather than isolated components. Cellular communication, a hallmark of evolutionary history, provides insights into how networks of cells create emergent phenomena.

Torday (2015) emphasized that evolutionary biology reflects the integration of molecular and cellular communication. This integration aligns with Chalmers' notion of a unified approach to consciousness, where subjective experience emerges from the coordination of physiological processes. By linking the synchronic (moment-to-moment) and diachronic (evolutionary) aspects of cellular life, holism bridges the gap between physical mechanisms and conscious experience.

## *Applications of Holism*

The principles of holism extend beyond theoretical insights into consciousness, offering practical solutions for challenges in medicine, environmental science, and artificial intelligence (AI). In medicine, understanding consciousness as an emergent property of cellular networks informs approaches to treating neurodegenerative diseases and mental health disorders. By targeting cellular communication and homeostasis, therapeutic strategies can address the root causes of these conditions rather than their symptoms.

In environmental science, holism underscores the interconnectedness of ecosystems and their impact on human health. The Gaia hypothesis, which views Earth as a self-regulating system, aligns with the principles of cellular homeostasis in maintaining planetary balance (Wilson, 1998). Recognizing these parallels fosters a deeper understanding of the relationship between environmental sustainability and human well-being.

In AI, holism provides a framework for developing systems that emulate the complexity of biological networks. By incorporating principles of cellular communication and homeostasis, AI models can achieve greater adaptability and resilience, mirroring the emergent properties of consciousness (Torday, 2024).

## *Conclusion*

Holism and homeostasis are central to understanding the evolution of consciousness, offering insights into its biological origins and broader applications. By framing consciousness as an emergent property of interconnected systems, this perspective bridges the gap between evolutionary biology and philosophy. Holism not only illuminates the complexities of consciousness but also provides practical tools for addressing pressing challenges in medicine, environmental science, and AI, reinforcing its relevance across disciplines.

## References

Ben Maamar, M., et al. (2021). Epigenetic inheritance and the environment: Mechanisms and implications. *Nature Reviews Genetics, 22*(9), 547–561.

Berner, R. A., et al. (2007). The origin of high metabolic rates in terrestrial vertebrates. *Proceedings of the National Academy of Sciences, 104*(2), 548–554.

Bing-Canar, H., et al. (2016). Breathing practices and their effects on stress hormones: A review. *Frontiers in Psychology, 7*, 774.

Chalmers, D. J. (1995). Facing up to the problem of consciousness. *Journal of Consciousness Studies, 2*(3), 200–219.

Clack, J. A. (2012). *Gaining ground: The origin and evolution ofTetrapods*. Indiana University Press.

Clemens, T. L., et al. (2001). Parathyroid hormone-related protein is essential for skeletal development and pulmonary function. *Development, 128*(8), 1443–1450.
Diamond, J. (1999). The third Chimpanzee: The evolution and future of the human animal. .
Gould, S. J., & Vrba, E. S. (1982). Exaptation—A missing term in the science of form. *Paleobiology, 8*(1), 4–15.
Gray, M. W. (2017). Lynn Margulis and the endosymbiont hypothesis: 50 years later. *Molecular Biology and Evolution, 34*(6), 1407–1409.
Grof, S. (2010). *The Holotropic mind: The three levels of human consciousness and how they shape our lives*. HarperOne.
Maturana, H. R., & Varela, F. J. (1972). *Autopoiesis and cognition: The realization of the living*. Reidel Publishing.
Rodman, P. S., & McHenry, H. M. (1980). Bioenergetics and the origin of bipedalism. *American Journal of Physical Anthropology, 52*(1), 103–106.
Rubin, H., et al. (2004). Hypoxia-induced gene expression in mammals. *Physiological Reviews, 84*(1), 1–32.
Sagan, L. (1967). On the origin of mitosing cells. *Journal of Theoretical Biology, 14*(3), 255–274.
Torday, J. S. (2015). A central theory of biology. *Medical Hypotheses, 85*, 49–57.
Torday, J. S. (2019). The singularity of nature. *Progress in Biophysics and Molecular Biology, 142*, 23–31.
Torday, J. S. (2023). Consciousness, embodied quantum entanglement. *Progress in Biophysics and Molecular Biology, 177*, 125–128.
Torday, J. S. (2024). *Consciousness-based evolution*. CRC Press.
Torday, J. S., & Rehan, V. K. (2003). Mechanotransduction determines the structure and function of lung and bone: A theoretical model for the pathophysiology of chronic disease. *Cell Biochemistry and Biophysics, 37*(3), 235–246.
Weibel, E. R., et al. (1998). *Symmorphosis: On form and function in shaping life*. Harvard University Press.
Wilson, E. O. (1998). *Consilience: The unity of knowledge*. Vintage Books.
Xu, S., et al. (2018). The role of FOXP2 in the evolution of speech and language. *Annual Review of Genomics and Human Genetics, 19*, 59–80.

# Chapter 16
# Gödel's Logic and the Role of Consciousness

The quest to understand consciousness intersects profoundly with the principles of logic and mathematics. Among the most intriguing frameworks for exploring this connection is Kurt Gödel's incompleteness theorems, which challenge the completeness of formal systems and reveal the inherent limits of mechanistic reasoning. In this chapter, we delve into Gödel's logic to examine its implications for consciousness, highlighting the interplay between mathematical structures, biological systems, and subjective experience.

Gödel's incompleteness theorems demonstrate that no formal system can fully account for its own axioms, suggesting that understanding reality—whether in mathematics, biology, or consciousness—requires stepping beyond reductionist frameworks. This insight mirrors the challenges of grasping consciousness: a phenomenon that resists full explanation through mechanistic models. By integrating Gödel's principles with evolutionary biology, this chapter reframes consciousness as an emergent, self-referential property of life.

The role of consciousness in transcending logical limitations becomes particularly salient when considered alongside biological processes. Evolution, like mathematics, operates within systems defined by internal rules yet achieves adaptability and creativity that defy deterministic constraints. Cellular evolution, quantum phenomena, and nonlocal dynamics serve as examples of nature's capacity to exceed its own parameters, echoing Gödel's philosophical implications.

This chapter explores how Gödel's theorems inform our understanding of consciousness through five key sections. First, we outline Gödel's incompleteness theorems and their philosophical resonance with biology and cognition. Next, we examine how cellular processes and evolutionary adaptation embody Gödelian principles of self-reference and incompleteness. We then delve into the relationship between mathematics, logic, and nonlocal consciousness, highlighting parallels between Gödel's work and quantum mechanics. The chapter concludes by synthesizing these insights, proposing a unified framework that links Gödelian logic to the biology of consciousness.

 
J. S. Torday, R. G. Sacco, *Conscious Life*,
https://doi.org/10.1007/978-3-031-80500-4_16

Through this exploration, we aim to demonstrate that Gödel's incompleteness theorems are not merely abstract mathematical ideas but profound metaphors for understanding life and consciousness. By embracing the interplay between logic, biology, and self-awareness, we advance toward a holistic framework for unraveling the mystery of consciousness itself.

## Gödel's Incompleteness and the Role of Nonlocal Consciousness

Kurt Gödel's incompleteness theorems, introduced in 1931, underscore the inherent limitations of formal systems, demonstrating that any sufficiently complex mathematical framework contains truths that cannot be proven within its own rules. While Gödel's insights revolutionized mathematics and logic, they also resonate profoundly with questions about the nature of consciousness. This section explores how Gödel's incompleteness can be reconciled through the concept of nonlocal consciousness, linking mathematical principles to the universal and integrative properties of life.

### *Gödel's Theorems and the Boundaries of Formal Systems*

Gödel's first incompleteness theorem asserts that no formal system capable of arithmetic can be both consistent and complete; there will always be true statements that cannot be proven within the system. The second theorem extends this by showing that no system can demonstrate its own consistency. These principles highlight the self-referential limitations of closed systems, suggesting that understanding certain truths requires stepping outside the confines of the system itself (Gödel, 1931).

In the context of consciousness, Gödel's theorems invite us to consider whether human cognition—often modeled mechanistically—may likewise transcend its formal structures. This self-referential quality aligns with the concept of nonlocal consciousness, where awareness operates beyond localized, reductionist frameworks. Nonlocal consciousness, in this sense, completes Gödel's framework by introducing a meta-system: a holistic integration of phenomena that transcends formal limitations.

### *Parallels Between Gödel's Theorems and Bohm's Implicate Order*

David Bohm's concept of the Implicate Order provides a philosophical framework that complements Gödel's insights. Bohm proposed that the universe operates on two levels: the Explicate Order, representing the observable, structured reality, and the Implicate Order, an underlying, interconnected foundation that enfolds all potential states (Bohm, 1980). Gödel's incompleteness aligns with this view, as both highlight the necessity of an external, unifying principle to reconcile limitations within formal systems.

Consciousness, when viewed through Bohm's lens, emerges as a bridge between the Implicate and Explicate Orders. It functions as a nonlocal integrative force, connecting disparate phenomena into a coherent whole. Gödel's insights suggest that consciousness itself may be the "missing component" that resolves the incompleteness of formal systems, operating not as an external observer but as a participant in the system's unfolding complexity.

### *The Biological Basis of Nonlocal Consciousness*

While Gödel and Bohm provide the philosophical foundation, biology offers empirical support for nonlocal consciousness. Symbiogenesis, the process by which distinct organisms merge to form new, integrated entities, exemplifies how life transcends individual systems to achieve greater unity. For example, the incorporation of mitochondria into eukaryotic cells represents not only a physical merger but also a transformation that integrates energy regulation and genetic expression (Sagan, 1967).

This cellular-level phenomenon mirrors Gödel's notion of transcending boundaries, as symbiogenesis integrates disparate systems into a cohesive whole. Furthermore, the recursive relationship between cellular processes and universal laws—such as energy flow and entropy regulation—supports the idea that consciousness emerges as an extension of these integrative mechanisms (Torday, 2023).

### *Conclusion*

Gödel's incompleteness theorems, far from presenting an unsolvable paradox, open the door to a more profound understanding of consciousness. By introducing the concept of nonlocal consciousness, we see how the limitations of formal systems can be resolved through holistic integration. Whether framed through Bohm's Implicate Order, symbiogenesis, or universal laws, the interplay between mathematics, biology, and consciousness suggests that understanding reality requires embracing both local structures and their nonlocal, interconnected foundations.

## Cellular Physiology and the Continuum of Mathematics

The connection between cellular evolution and mathematical structures provides profound insights into the nature of life and its trajectory. Cellular physiology, which underpins adaptation and evolution, operates through patterns and processes that resonate with fundamental mathematical principles. This section explores the continuum between cellular memory, epigenetic inheritance, and the mathematical frameworks that mirror and guide these processes.

## *Epigenetic Inheritance and Mathematical Frameworks*

Epigenetic inheritance serves as a mechanism through which cellular memory is extended across generations. This process involves chemical modifications to DNA that do not alter the genetic code but influence gene expression, allowing cells to adapt to environmental changes in real time and pass these adaptations to offspring. This iterative process of encoding environmental information aligns with the recursive and self-referential properties of Rowlands' Rewrite Mathematics (Rowlands, 2015).

In Rowlands' framework, zero serves as the attractor around which transformations occur, mirroring how epigenetic modifications preserve homeostasis while enabling evolutionary innovation. Similarly, Kauffman's Knot Theory provides a metaphor for cellular processes, with the tying and untying of knots reflecting the dynamic nature of adaptation. Cellular systems can be viewed as "knotted" networks of interactions that maintain structural integrity while accommodating external perturbations (Kauffman, 2006).

## *Fibonacci and the Periodic Table: Synchronic and Diachronic Dimensions*

The Fibonacci sequence, a mathematical series where each term is the sum of the two preceding terms, has long been recognized as a representation of natural patterns. From the arrangement of leaves on a stem to the spirals of a nautilus shell, Fibonacci numbers reflect growth processes that integrate past states into future developments. Cellular evolution embodies this principle, where synchronic (moment-to-moment) processes, such as metabolic regulation, coexist with diachronic (historical) processes, such as genetic inheritance and epigenetic marks (Livio, 2002).

The Periodic Table of Elements offers another example of mathematical homology. Its synchronic structure, represented by horizontal rows, reflects elemental similarities, while its diachronic relationships, evident in diagonal trends, illustrate evolutionary transitions in atomic behavior. This dual perspective mirrors cellular evolution, where immediate physiological needs and long-term adaptive strategies coexist. By linking the behavior of atoms to the behavior of cells, the periodic table serves as a bridge between the microscopic and cosmic scales of evolution (Torday, 2024).

### *Iterative Processes and Mathematical Homologies*

Life's trajectory is shaped by iterative processes that continuously refine and optimize cellular and organismal functions. Mathematical homologies provide a framework for understanding these processes. For example, the Fibonacci sequence reflects the balance between growth and resource allocation, while Knot Theory highlights the resilience and adaptability of interconnected systems. Cellular adaptation, driven by environmental feedback and genetic mechanisms, follows a similarly recursive and self-organizing pattern.

Mathematical structures like these not only describe the principles governing life but also guide its trajectory. They demonstrate that evolution is not random but deeply connected to universal patterns of order and complexity. Cellular adaptation, whether through epigenetic inheritance or metabolic regulation, reflects an underlying mathematical logic that integrates environmental challenges into the evolutionary narrative.

### *Conclusion*

By linking cellular physiology to mathematical frameworks, we gain a deeper understanding of life's adaptive mechanisms. Epigenetic inheritance, the Fibonacci sequence, and the periodic table reveal the synchronic and diachronic dimensions of evolution, illustrating how mathematical principles shape biological processes. These insights underscore the unity of biology and mathematics, offering a continuum that bridges cellular and cosmic evolution.

## Gravity, Symbiogenesis, and the Unity of the Cosmos

The interplay between gravity, symbiogenesis, and universal laws illuminates the interconnectedness of life and the cosmos. Gravity, as a fundamental force, shapes both the physical structure of the universe and the physiological processes within living systems. Simultaneously, symbiogenesis integrates environmental challenges into the fabric of cellular physiology, creating a dynamic alignment with cosmic principles. This section explores how these forces have guided the evolution of life and reflect the unity of biological and physical laws.

## *Gravity: A Driver of Cellular Differentiation and Evolutionary Adaptation*

Gravity is an omnipresent force that has shaped life since its inception.

At the cellular level, gravity influences differentiation and organization, guiding the spatial distribution of organelles and the structural integrity of tissues. During the evolutionary transition from aquatic to terrestrial environments, gravity played a pivotal role in vertebrate adaptation. The increased gravitational force on land necessitated significant physiological changes, including skeletal stiffening and lung alveolarization, to support mobility and efficient oxygen exchange (Clack, 2012). These adaptations highlight gravity's role as a driver of physiological innovation, pushing organisms to evolve structural and metabolic solutions for survival.

Microgravity experiments provide additional evidence of gravity's influence on cellular processes. Studies show that in microgravity, cells lose their phenotypic identity, reverting to a more primitive state (Torday, 2003). This phenomenon underscores the dependence of cellular differentiation on gravitational cues, linking environmental forces to developmental pathways. These findings suggest that gravity is not merely a physical constraint but an essential factor in maintaining cellular organization and homeostasis, influencing life's trajectory from unicellular organisms to complex multicellular systems.

## *Symbiogenesis: Integrating Environmental Challenges into Physiology*

Symbiogenesis, the process by which two or more distinct organisms merge to form a new entity, serves as a mechanism for integrating environmental challenges into physiology. This evolutionary strategy allows life to assimilate external pressures, transforming them into internal adaptations that enhance survival and complexity. For instance, the integration of mitochondria into eukaryotic cells exemplifies how symbiogenesis facilitated energy efficiency and metabolic innovation, laying the groundwork for complex life forms (Sagan, 1967).

In the context of gravity, symbiogenesis enabled organisms to align their physiology with gravitational forces. For example, during the transition from water to land, cellular systems adapted to new environmental pressures, such as reduced buoyancy and increased oxygen availability, by developing novel structures and functions. These adaptations, mediated by symbiogenetic processes, illustrate how life evolves in response to cosmic principles, embedding environmental challenges into its biological framework.

### *Cellular Processes and the Homology with Cosmic Phenomena*

Cellular processes reflect the fundamental laws of physics, drawing parallels between biological systems and cosmic phenomena. The cellular membrane, for example, can be likened to the event horizon of a black hole, a boundary that separates and regulates the flow of information and energy. Similarly, gravitational distortion, as described by Einstein's theory of general relativity, mirrors the deformation and reorganization of cellular structures under stress (Einstein, 1916). These homologies underscore the unity of physical and biological principles, revealing a shared framework that governs both the cosmos and life.

Gravity's role in shaping the curvature of spacetime finds a biological parallel in the way cells respond to mechanical forces. Cellular mechanotransduction, the process by which cells convert mechanical stimuli into biochemical signals, reflects this relationship. Just as gravity molds the structure of the universe, it influences cellular architecture and function, reinforcing the connection between macrocosmic and microcosmic systems.

### *Conclusion*

The interplay between gravity, symbiogenesis, and universal laws underscores the unity of life and the cosmos. Gravity serves as a driver of cellular differentiation and evolutionary adaptation, while symbiogenesis integrates environmental challenges into the physiological fabric of life. Together, these forces reveal that cellular processes are not isolated phenomena but reflections of the broader physical laws that govern the universe. By examining these connections, we gain a deeper understanding of how life evolves in harmony with cosmic principles, emphasizing the universality of biological and physical systems.

## Gödel's Theorems Revisited Through Cellular and Quantum Frameworks

Gödel's incompleteness theorems have long illuminated the limitations of formal mathematical systems, revealing that no logical framework can be both complete and self-consistent. However, when reevaluated through the lens of biology and quantum mechanics, Gödel's work finds new dimensions that extend its implications. This perspective suggests that nonlocal consciousness, quantum phenomena, and cellular biology collectively bridge the gaps Gödel identified, completing the mathematical framework in ways that classical physics and logic alone cannot achieve.

## *Nonlocal Consciousness and Gödel's Theorems*

Gödel's theorems highlight the inherent limits of closed logical systems, where certain truths remain unprovable within the system itself (Gödel, 1931). Nonlocal consciousness, as a phenomenon that transcends spatial and temporal boundaries, provides a way to address these gaps. Quantum mechanics demonstrates nonlocality through phenomena like quantum entanglement, where particles remain interconnected regardless of distance (Penrose, 1996). Similarly, consciousness appears to operate on multiple levels—local, as seen in individual neural processing, and nonlocal, as reflected in phenomena like intuition or shared awareness.

Physiologically, this dual nature of consciousness is observable in microgravity experiments. Cells exposed to microgravity revert to a more primitive state, illustrating their dependency on both local gravitational cues and nonlocal adaptive mechanisms for maintaining structure and function (Torday, 2003). These findings suggest that consciousness operates on a two-tiered system: a local component rooted in cellular processes and a nonlocal component integrated into broader cosmic dynamics. Together, these layers form a continuum that mirrors Gödel's interplay between provable and unprovable truths, bridging the gaps in mathematical logic.

## *The Möbius Strip: A Model for Cellular and Conscious Unity*

The cell, as the foundational unit of life, offers an intriguing parallel to Gödel's mathematical insights. Framed as a Möbius strip—a surface with only one side and one edge—the cell encapsulates the unity of the observer and the observed (Torday, 2021). This structure reflects the seamless integration of internal and external environments, where cellular processes adapt dynamically to both local stimuli and nonlocal influences.

The Möbius strip metaphor also extends to consciousness, where the act of observation influences the observed, as seen in quantum mechanics. The double-slit experiment demonstrates how the behavior of particles changes based on observation, a phenomenon that aligns with the cellular integration of internal and external information. In this framework, consciousness becomes both the observer and the mechanism by which the universe observes itself, tying Gödel's logic to the principles of quantum and cellular unity.

## *Bridging Biology, Mathematics, and Quantum Mechanics*

By integrating the perspectives of biology, mathematics, and quantum mechanics, we can reinterpret Gödel's theorems as part of a larger framework that includes life and consciousness. The self-referential nature of Gödel's logic mirrors the feedback loops inherent in biological systems, where cellular memory and adaptation reflect iterative, self-correcting processes (Tegmark, 2014). Similarly, the homologies between mathematical principles like the Fibonacci sequence and cellular evolution reveal a synchronicity that aligns life with universal laws.

This synthesis suggests that the limitations Gödel identified in mathematical systems may be artifacts of an incomplete framework—one that requires the inclusion of nonlocal consciousness and quantum phenomena to achieve coherence. By incorporating these elements, Gödel's theorems become not just an endpoint but a gateway to understanding the unity of life, mathematics, and the cosmos.

### *Conclusion*

Gödel's incompleteness theorems, when viewed through the lens of biology and quantum mechanics, reveal profound insights into the nature of life and consciousness. Nonlocal phenomena, as observed in quantum mechanics and cellular processes, provide the missing components to complete Gödel's logical framework. The cell, as a Möbius strip-like entity, exemplifies the unity of the observer and the observed, bridging the divide between local and nonlocal properties. This reevaluation highlights the interconnectedness of mathematical, biological, and quantum principles, offering a holistic perspective on the fundamental nature of existence.

## Implications for Science, Philosophy, and Consciousness Studies

Gödel's incompleteness theorems have long been celebrated for revealing the inherent limitations of formal systems, suggesting that no mathematical framework can encapsulate all truths within its bounds. However, these limitations can also be interpreted as an invitation to expand our understanding of the universe by integrating consciousness, a phenomenon that transcends the boundaries of reductionist science. This section explores how Gödel's work, when viewed through the lens of nonlocal consciousness, provides a bridge between mathematics, biology, and cosmology, offering a unified framework for understanding life and the universe.

### *Gödel's Theorems and the Necessity of Consciousness in Science*

Philosophically, Gödel's work implies that formal systems are inherently incomplete without external input. This notion aligns with arguments by philosophers like Lucas (1961), who posited that human consciousness operates beyond the capabilities of formal logic. Consciousness, particularly its nonlocal aspects, introduces an element of universality and interconnectedness that formal systems cannot achieve alone.

This idea resonates with the work of Goldstein (2006), who argued that the limitations of mathematics highlight the need for holistic perspectives that include subjective experience. Nonlocal consciousness, as evidenced by quantum entanglement and biological phenomena such as cellular memory, embodies this holistic dimension. By incorporating consciousness into scientific frameworks, we move closer to addressing the epistemological gaps identified by Gödel, creating a more comprehensive understanding of reality.

## *Nonlocal Consciousness and a New Scientific Paradigm*

Nonlocal consciousness offers a powerful framework for uniting disparate fields of inquiry. For example, in biology, cellular processes such as epigenetic inheritance demonstrate how information can be transmitted across generations, echoing the continuity of mathematical and cosmic laws (Morowitz, 2004). Similarly, in cosmology, the fine-tuning problem—the apparent precision with which physical constants are set to allow life—suggests a deeper integration between consciousness and universal laws.

Torday (2024) has proposed a paradigm that merges biology, mathematics, and cosmology into a single continuum, where life is not an isolated phenomenon but a reflection of the universe's intrinsic logic. In this view, Gödel's incompleteness theorems serve as a metaphor for the necessity of interdisciplinary approaches that include consciousness as a central element. By linking quantum phenomena, cellular processes, and cosmological principles, this paradigm reveals the interconnected nature of all systems.

## *Completing the Universal Framework*

Gödel's theorems, when reinterpreted through the lens of nonlocal consciousness, suggest that the universe itself operates as a continuum of laws and life. This perspective positions consciousness not as an emergent property of complex systems but as a fundamental aspect of the cosmos, woven into its very fabric. The integration of formal systems with nonlocal phenomena allows for a complete understanding of the universe, bridging the gaps between mathematics, physics, and biology.

This unified framework has profound implications for consciousness studies, offering new insights into the nature of awareness, free will, and the relationship between the observer and the observed. It also challenges traditional scientific paradigms, calling for a shift from reductionist methodologies to holistic approaches that embrace the complexity and interconnectedness of life.

## *Conclusion*

Gödel's incompleteness theorems highlight the necessity of integrating consciousness into scientific frameworks, offering a new perspective on the unity of life and the universe. By incorporating nonlocal consciousness, we can address the limitations of formal systems and create a comprehensive understanding that spans mathematics, biology, and cosmology. This unified framework not only resolves the philosophical challenges posed by Gödel's work but also opens new avenues for interdisciplinary research, positioning consciousness as the bridge between the known and the unknown.

## References

Bohm, D. (1980). *Wholeness and the implicate order*. Routledge.
Clack, J. A. (2012). *Gaining ground: The origin and evolution of Tetrapods*. Indiana University Press.
Einstein, A. (1916). The foundation of the general theory of relativity. *Annalen der Physik, 49*(7), 769–822.
Gödel, K. (1931). On formally undecidable propositions of principia Mathematica and related systems I. *Monatshefte für Mathematik und Physik, 38*(1), 173–198.
Goldstein, R. (2006). *Incompleteness: The proof and Paradox of Kurt Gödel*. Norton.
Kauffman, L. H. (2006). *Knot theory and physics*. World Scientific Publishing.
Livio, M. (2002). *The Golden ratio: The story of Phi, the world's most astonishing number*. Broadway Books.
Lucas, J. R. (1961). Minds, machines, and Gödel. *Philosophy, 36*(137), 112–127.
Morowitz, H. J. (2004). *The emergence of everything: How the world became complex*. Oxford University Press.
Penrose, R. (1996). *Shadows of the mind: A search for the missing science of consciousness*. Oxford University Press.
Rowlands, P. (2015). *Zero to infinity: The foundations of physics*. World Scientific Publishing.
Sagan, L. (1967). On the origin of mitosing cells. *Journal of Theoretical Biology, 14*(3), 255–274.
Tegmark, M. (2014). *Our mathematical universe: My quest for the ultimate nature of reality*. Knopf.
Torday, J. S. (2003). Parathyroid hormone-related protein is a gravisensor in lung and bone cell biology. *Advances in Space Research, 32*, 1569–1576.
Torday, J. S. (2021). Life is a Möbius strip. *Progress in Biophysics and Molecular Biology, 167*, 41–45.
Torday, J. S. (2023). The holism of cosmology and consciousness. *Progress in Biophysics and Molecular Biology, 177*, 181–184.
Torday, J. S. (2024). Symbiogenesis predicts the monism of the cosmos. *Progress in Biophysics and Molecular Biology, 191*, 58–62.

# Part V
# Toward a Unified Vision of Life and the Cosmos

# Chapter 17
# Guiding Principles of Cellular Evolution

The story of life is rooted in the fundamental processes of cellular evolution—a tapestry woven from interactions between environmental forces, molecular adaptations, and the dynamic flow of energy and information. This chapter explores the guiding principles that have shaped the trajectory of cellular evolution, emphasizing its role as the foundation of biological complexity and a key to understanding the interconnectedness of life.

At the heart of cellular evolution lies the interplay between environmental pressures and adaptive responses. From the primordial micelles that formed the first cellular boundaries to the sophisticated multicellular organisms we observe today, evolution has been guided by principles such as homeostasis, energy efficiency, and information storage. These processes reveal a profound unity between physics, chemistry, and biology, highlighting the universality of life's mechanisms across scales.

In this chapter, we examine cellular evolution through the lens of symbiogenesis, epigenetics, and systemic integration. Symbiogenesis emerges as a central mechanism for assimilating environmental challenges into physiology, providing the foundation for innovation and resilience. Similarly, epigenetic inheritance underscores how life adapts across generations, preserving evolutionary memory while remaining flexible to change. Finally, systemic integration links cellular processes to broader biological and cosmological contexts, demonstrating how life thrives by uniting local and global dynamics.

The guiding principles of cellular evolution also extend into broader philosophical and practical domains. Understanding these principles enables us to address pressing challenges in medicine, ecology, and technology, offering insights into chronic diseases, environmental sustainability, and even artificial intelligence. Moreover, cellular evolution provides a framework for rethinking the relationship between humanity and the natural world, encouraging a holistic approach that values interconnectedness and adaptability.



J. S. Torday, R. G. Sacco, *Conscious Life*,
https://doi.org/10.1007/978-3-031-80500-4_17

This chapter invites readers to explore the foundational mechanisms driving cellular evolution and their implications for life, consciousness, and the cosmos. By tracing the guiding principles that shape life's trajectory, we uncover the profound wisdom embedded in the cellular processes that connect us to our evolutionary past and guide us toward a sustainable future.

## The Cell as a Conscious Entity

The foundational premise that cells possess a form of intrinsic consciousness has far-reaching implications for understanding life's evolution and interconnectedness. Rather than viewing consciousness as an exclusive trait of complex organisms, this perspective positions it as an emergent property originating from cellular processes. Evidence from experimental biology, evolutionary theory, and molecular studies supports the notion that cells are not merely biochemical automata but entities capable of perceiving, responding to, and integrating environmental challenges.

One compelling line of evidence for cellular consciousness comes from studies conducted in microgravity, which demonstrate how gravitational forces play a critical role in stabilizing cellular phenotypes and preserving evolutionary memory. In microgravity environments, cells lose their differentiated phenotypic states, reverting to a more primitive and undifferentiated condition. This reversion highlights the cell's dependency on external forces like gravity to maintain homeostasis and specialized functions. These findings suggest that cells possess a form of memory tied to their evolutionary history, enabling them to adapt dynamically to external conditions while maintaining an internal sense of continuity (Torday, 2023; Camirand et al., 2016).

Symbiogenesis provides another compelling mechanism through which cells exhibit a form of consciousness. By assimilating environmental challenges into their physiology, cells do more than merely survive—they innovate. Symbiogenesis occurs when cells integrate external elements, such as endosymbiotic bacteria, into their structure, transforming environmental pressures into evolutionary advantages. This process reflects a kind of cellular awareness, where challenges are not only detected but also are incorporated into the system to ensure survival and functional enhancement. Such adaptive processes showcase the cell's capacity for holistic integration, a trait often associated with consciousness.

The concept of cellular memory further reinforces the idea of the cell as a conscious entity. Cellular memory refers to the ability of cells to store information about their interactions with the environment, enabling them to reference past experiences and inform future responses. This phenomenon is evident in epigenetic inheritance, where environmental influences are encoded in genetic material and transmitted across generations. Studies on unicellular organisms reveal that even in their simplest forms, cells operate as holistic systems capable of referencing their evolutionary history. This intrinsic memory allows unicellular organisms to act as coherent systems that balance survival with adaptation (Baluska & Reber, 2019).

In this light, the unicellular state is not just a precursor to multicellular complexity but a self-contained system embodying the principles of homeostasis, adaptation, and awareness. By integrating environmental inputs, maintaining internal stability, and storing evolutionary information, cells demonstrate a capacity for decision-making and intentionality. These attributes align with a broader definition of consciousness that extends beyond cognition to encompass any system capable of sensing, responding to, and adapting within its environment.

Understanding the cell as a conscious entity invites a paradigm shift in biology, emphasizing the continuity of life and the interconnectedness of all living systems. This perspective not only bridges the gap between physiology and philosophy but also redefines the origins of consciousness as a fundamental property of life itself.

## The Origin of Life and the Role of Symbiogenesis

The emergence of life on Earth, beginning with the first cells, marks a pivotal intersection of cosmology, chemistry, and biology. At the heart of this origin story lies the concept of symbiogenesis, a transformative process in which existential challenges are not merely confronted but assimilated into the evolving physiology of life. Symbiogenesis serves as a key driver of evolutionary complexity and adaptability, underpinning the development of life as we know it.

The origins of life are rooted in the unique physicochemical properties of lipid molecules, which were carried to Earth by asteroids and comets. These lipids spontaneously formed micelles—spherical aggregates of molecules with a hydrophilic exterior and a hydrophobic interior—when exposed to water. This self-organization of lipids is considered the first step toward the formation of protocells, which provided the structural basis for life. These micelles acted as semipermeable membranes, enabling the compartmentalization of chemical reactions essential for maintaining homeostasis and energy flow (Groen et al., 2012). This process not only bridged abiotic and biotic systems but also laid the groundwork for evolution by establishing the physical boundaries necessary for the emergence of life.

The formation of micelles represents a "quantum leap" in the origin of life, linking their emergence to quantum mechanics and the foundational principles of evolution. The alignment of lipid molecules into organized structures mirrors the probabilistic yet deterministic nature of quantum processes, such as entanglement. As micelles formed and began to sustain localized chemical reactions, they embodied the principles of quantum coherence—where discrete components function cohesively as a system. This quantum perspective underscores the transition from inert matter to dynamic, living systems and highlights the profound relationship between physics and biology in the evolutionary narrative (Classen & Spooner, 1996; Torday, 2022).

Symbiogenesis, the process by which organisms integrate external challenges into their physiology, further illuminates the pathway from simple protocells to complex life forms. Initially, environmental stressors posed existential threats to early life. However, rather than merely resisting these threats, early cells absorbed and incorporated them, transforming adversity into adaptive advantage. For instance, the integration of bacterial endosymbionts into early eukaryotic cells led to the formation of mitochondria and chloroplasts, enabling more efficient energy production and photosynthesis. This capacity to internalize and repurpose environmental challenges is the hallmark of symbiogenesis and the basis for the exponential complexity observed in evolutionary history (Sagan, 1967).

The evolutionary trajectory of life, from lipid micelles to multicellular organisms, is therefore deeply intertwined with the principles of self-organization and adaptability. Symbiogenesis not only enabled the survival of early cells in a hostile environment but also fostered the development of increasingly intricate biological systems. By assimilating existential challenges, cells created a blueprint for evolutionary innovation, demonstrating that the essence of life lies in its capacity to transform threats into opportunities for growth.

In this context, the origins of life can be viewed as a continuum of increasing complexity and coherence, where the interplay of physical laws, chemical properties, and biological mechanisms drives evolution forward. This narrative positions symbiogenesis as a fundamental principle of life, reflecting the unity of the cosmos and the cellular processes that sustain it.

## Mathematics as the Language of Cellular Evolution

Mathematics serves as a universal language that transcends disciplines, offering profound insights into the intricate mechanisms of cellular evolution. From the Fibonacci sequence to the Möbius strip, mathematical principles illuminate the dynamic interplay of cellular processes, highlighting how life continuously adapts and evolves across generations. This section explores the application of mathematical frameworks to cellular and evolutionary dynamics, positioning them as foundational elements of life's trajectory.

The Fibonacci sequence, a mathematical pattern where each number is the sum of the two preceding ones, provides a compelling metaphor for understanding evolutionary biology. This sequence reflects the iterative and adaptive nature of cellular processes, where the past informs the present to shape the future. For instance, the spatial arrangement of molecules within cellular structures often follows Fibonacci-like patterns, optimizing functionality and efficiency. Similarly, the Möbius strip, a one-sided surface, represents the seamless continuity between internal and external cellular environments, symbolizing the unification of life's synchronic (moment-to-moment) and diachronic (historical) dimensions (Tegmark, 2014; Torday, 2021). These mathematical constructs underscore the inherent order and coherence within biological systems, bridging the microscopic and macroscopic scales of evolution.

Knot Theory and Peter Rowlands' Rewrite Mathematics further enrich our understanding of cellular replication and development. Knot Theory, which studies the mathematical properties of tangled and untangled loops, offers a powerful analogy for the folding and unfolding processes in DNA replication and protein synthesis. Just as knots can be untied and restructured, cellular mechanisms exhibit a similar ability to reorganize and adapt to environmental challenges. Rowlands' Rewrite Mathematics, which emphasizes zero as an attractor and a point of symmetry, mirrors the cellular tendency to reset and realign during replication and repair processes. These mathematical frameworks reveal how cellular systems maintain integrity and adaptability while navigating the complexities of life (Kauffman, 2006; Rowlands, 2015).

Epigenetic inheritance adds another layer of mathematical sophistication to cellular evolution. This process, where environmental influences leave reversible "marks" on DNA, enables organisms to transmit adaptive information to subsequent generations. Epigenetics functions as a Fibonacci-like mechanism, preserving cellular memory while introducing necessary adaptations to meet environmental demands. For example, epigenetic modifications allow organisms to optimize metabolic pathways, regulate gene expression, and enhance resilience against stressors, ensuring the survival and evolution of life forms (Ben Maamar et al., 2021). This cyclical interplay between preservation and adaptation reflects the mathematical elegance of biological systems.

The convergence of these mathematical principles highlights the profound unity between life's cellular processes and the universal laws of mathematics. The Fibonacci sequence, Möbius strip, Knot Theory, and Rewrite Mathematics collectively offer a lens through which we can decode the logic of evolution. They reveal how life's trajectory is not merely a series of random mutations but a deliberate and harmonious dance guided by mathematical order and adaptability.

Through these mathematical frameworks, we can better understand the essence of cellular evolution as a process that seamlessly integrates past experiences, present conditions, and future possibilities. This perspective not only deepens our appreciation for the complexity of life but also underscores the interconnectedness of mathematics, biology, and the cosmos as a unified continuum.

## The Water-Land Transition and Bipedalism

The evolutionary transition from water to land marked a transformative period in the history of life, driving adaptations that fundamentally reshaped physiology and morphology. This transition underscores the profound interconnectedness of environmental challenges, cellular responses, and evolutionary trajectories. Hypoxia and gravitational forces, in particular, played pivotal roles in shaping the adaptations necessary for terrestrial life, including the development of lungs, endothermy, and bipedalism.

As organisms transitioned from aquatic environments to land, they encountered significant shifts in oxygen availability and the demands of gravity. Hypoxia, characterized by reduced oxygen levels, served as a critical stressor that drove the evolution of more efficient respiratory systems. The vertebrate lung, and specifically the evolution of the alveolus, exemplifies this adaptation. Alveoli, the microscopic sacs within the lungs, maximize oxygen exchange by increasing surface area relative to volume. This innovation was largely mediated by parathyroid hormone-related protein (PTHrP), a key signaling molecule that regulates alveolar formation and function (Torday & Rehan, 2012). By optimizing respiratory efficiency, alveolar evolution allowed vertebrates to thrive in oxygen-scarce terrestrial environments, supporting higher metabolic rates and the eventual development of endothermy.

Gravitational forces posed additional challenges during the water-to-land transition, necessitating structural adaptations to support body weight and maintain balance. Skeletal stiffening, achieved through the mineralization of cartilage into bone, provided the necessary rigidity and strength for terrestrial locomotion. This process was also influenced by PTHrP, which played a crucial role in coordinating skeletal development and lung adaptation. These interconnected physiological changes highlight the synergistic evolution of organ systems in response to environmental pressures (Rubin et al., 2004).

The emergence of bipedalism represents one of the most profound adaptations during this evolutionary period. By shifting to an upright posture, early terrestrial organisms freed their forelimbs for other functions, such as manipulation and tool use, setting the stage for the development of advanced cognitive abilities. Interestingly, bipedalism and the associated expansion of the central nervous system can be traced back to the original micelle, the protocell structure that laid the groundwork for life. This atavistic connection suggests that the transition from water to land was not merely a linear progression but a complex reorganization of evolutionary principles rooted in the earliest forms of life (Daeschler et al., 2006).

The water-to-land transition also catalyzed the evolution of endothermy, enabling organisms to regulate their internal body temperature independent of external conditions. This adaptation, driven by metabolic stress, allowed for greater activity levels and ecological flexibility. The integration of these physiological changes reflects the inherent adaptability of life, with each innovation building upon the foundational principles established by the protocell.

In conclusion, the water-to-land transition illustrates the dynamic interplay between environmental pressures and evolutionary responses. From the development of lungs and skeletal systems to the emergence of bipedalism and endothermy, this period of evolution underscores the resilience and adaptability of life. These adaptations, rooted in cellular processes and guided by environmental demands, reveal the profound continuity between the earliest life forms and the complexity of modern organisms.

## Consciousness as the Integration of Evolution and Cosmology

Consciousness, the most profound and enigmatic feature of life, emerges as a unified phenomenon that intertwines evolutionary biology and cosmology. This synthesis bridges cellular processes and universal dynamics, underscoring the interconnectedness of life and the cosmos. By examining the interplay between cellular homeostasis, quantum entanglement, and cosmological principles, a coherent framework for understanding consciousness begins to take shape.

At its core, consciousness is a product of cellular homeostasis—a mechanism by which cells maintain internal stability in response to external fluctuations. Cellular processes, such as energy regulation and signaling, create a self-sustaining system capable of adapting to environmental challenges. These processes reflect a broader cosmological order, where energy flows and entropy govern the dynamics of the universe. Quantum entanglement, which connects particles instantaneously across vast distances, mirrors the way cellular networks communicate and integrate information, suggesting that the emergent properties of consciousness are deeply rooted in quantum mechanics (Torday, 2024).

Carlo Rovelli's concept of "White Holes" provides a compelling analogy for understanding consciousness as a continuum between disequilibrium and stability. In cosmology, White Holes are theoretical constructs that eject matter and energy, contrasting the gravitational pull of Black Holes. Similarly, cellular memory operates as a dynamic system that maintains disequilibrium while enabling continuity. This disequilibrium is essential for life, allowing cells to respond to and integrate environmental changes over time (Rovelli, 2023). Cellular processes, such as epigenetic inheritance, exemplify this continuity by preserving evolutionary history while adapting to new conditions.

The integration of evolutionary and cosmological perspectives extends to the philosophical implications of consciousness. David Chalmers' "hard problem" of consciousness—how subjective experiences arise from physical processes—finds a potential resolution in the dynamic interplay between the local and the nonlocal. Cellular evolution, driven by homeostasis and quantum phenomena, provides a biological foundation for the emergence of subjective awareness. Simultaneously, the cosmos, governed by universal laws, offers the broader context within which life and consciousness evolve. This duality suggests that consciousness is not merely a byproduct of evolution but a fundamental property of the universe, inseparably connected to its underlying structure (Chalmers, 1995).

Buddhist scholar David Loy further reinforces this perspective by proposing that the interconnectedness of life and the cosmos reflects an underlying unity. Cellular evolution, which integrates environmental inputs into physiological adaptations, mirrors the philosophical notion of interdependence. This perspective aligns with the view that consciousness is a holistic phenomenon arising from the integration of biological and cosmological processes (Loy, 1988).

In synthesizing these insights, consciousness emerges as a bridge between the microcosm of cellular evolution and the macrocosm of the universe. This integrative framework has profound implications for science, philosophy, and humanity's understanding of its place in the cosmos. By recognizing the interconnected nature of life and the universe, a unified perspective on consciousness can illuminate its role as a driving force of existence, underscoring its profound significance in shaping the trajectory of life.

## References


Baluska, F., & Reber, A. (2019). Sentience and consciousness in single cells: How the first minds emerged in unicellular species. *BioEssays, 41*(10), e1800229.

Ben Maamar, M., Nilsson, E. E., & Skinner, M. K. (2021). Epigenetic transgenerational inheritance. *Biology of Reproduction, 105*, 570–592.

Camirand, A., et al. (2016). The role of parathyroid hormone-related protein in osteoblast response to microgravity. *PLoS One, 11*, e0160034.

Chalmers, D. J. (1995). Facing up to the problem of consciousness. *Journal of Consciousness Studies, 2*(3), 200–219.

Classen, D. E., & Spooner, B. S. (1996). Liposome formation in microgravity. *Advances in Space Research, 17*, 151–160.

Daeschler, E. B., Shubin, N. H., & Jenkins, F. A. (2006). A Devonian tetrapod-like fish and the evolution of the tetrapod body plan. *Nature, 440*(7085), 757–763.

Groen, J., et al. (2012). Polycyclic aromatic hydrocarbons as plausible prebiotic membrane components. *Origins of Life and Evolution of Biospheres, 42*, 295–306.

Kauffman, L. (2006). *Formal Knot theory*. Dover Publications.

Loy, D. (1988). *A study in comparative philosophy*. Yale University Press.

Rovelli, C. (2023). *White holes*. Riverhead Books.

Rowlands, P. (2015). *The foundations of physical law*. World Scientific, Singapore.

Rubin, L. P., et al. (2004). Arrested pulmonary alveolar cytodifferentiation. *Developmental Dynamics, 230*, 278–289.

Sagan, L. (1967). On the origin of mitosing cells. *Journal of Theoretical Biology, 14*(3), 255–274.

Tegmark, M. (2014). *Our mathematical universe*. Knopf.

Torday, J. S. (2021). Life is a Möbius strip. *Progress in Biophysics and Molecular Biology, 167*, 41–45.

Torday, J. S. (2022). Evolution, gravity, and the topology of consciousness. *Progress in Biophysics and Molecular Biology, 174*, 50–54.

Torday, J. S. (2023). Consciousness, embodied quantum entanglement. *Progress in Biophysics and Molecular Biology, 177*, 125–128.

Torday, J. S. (2024). The quantum cell. *Progress in Biophysics and Molecular Biology, 188*, 24–30.

Torday, J. S., & Rehan, V. K. (2012). *Evolutionary biology, cell-cell communication and complex disease*. Wiley.

# Chapter 18
# The Holistic Design of Biosystems

The intricate complexity of biosystems reflects a fundamental principle of nature: life is designed not as a collection of isolated parts but as an integrated whole. This chapter explores the holistic architecture of biosystems, emphasizing the interplay between cellular processes, physiological networks, and environmental interactions. By examining life through the lens of systems biology, evolution, and quantum mechanics, we can appreciate how biosystems are dynamically structured to adapt, evolve, and sustain themselves.

From the cellular level to entire ecosystems, biosystems are characterized by their ability to balance stability with adaptability—a dynamic equilibrium known as homeostasis. This balance emerges from the intricate coordination of energy flows, signaling pathways, and genetic regulation within and across organisms. At the core of this design lies a network of interdependencies that mirrors universal patterns of organization, where physical, chemical, and biological principles converge to create a unified system.

In this chapter, we delve into the foundational aspects of biosystem design, starting with the cellular level, where life's building blocks exhibit emergent properties that transcend their individual components. We explore the evolutionary mechanisms that have shaped biosystems over billions of years, demonstrating how symbiotic relationships and environmental challenges have driven their complexity. From the role of epigenetic inheritance to the integration of quantum principles, biosystems embody a seamless connection between local adaptation and universal laws.

Beyond the biological, the design of biosystems offers profound insights into the philosophy of life. It reveals how evolutionary processes have not only optimized survival but also integrated life into a broader cosmic narrative. This perspective challenges reductionist views, emphasizing the importance of holism in understanding the unity of life and its interconnectedness with the universe.

The chapter also examines the implications of biosystem design for medicine, sustainability, and the future of humanity. By embracing a holistic framework, we

 
J. S. Torday, R. G. Sacco, *Conscious Life*,
https://doi.org/10.1007/978-3-031-80500-4_18

can develop innovative approaches to health, environmental stewardship, and technological advancement that align with the principles of life itself.

Ultimately, the holistic design of biosystems is a testament to the elegance of life's evolutionary journey. It demonstrates that the seemingly disparate elements of biology, chemistry, physics, and philosophy are threads in a single tapestry—a dynamic, interconnected system that bridges the microcosm of the cell with the macrocosm of the cosmos. This chapter invites readers to reflect on the profound significance of this unity and its implications for our understanding of life and consciousness.

## Symbiogenesis as the Nexus of Life and Cosmos

Symbiogenesis, the integration of distinct biological entities to form a unified organism, serves as a powerful unifying principle connecting the realms of biology and cosmology. This evolutionary mechanism transcends the traditional framework of Darwinian selection, emphasizing cooperative relationships over competition. By framing symbiogenesis as a process that assimilates existential threats into physiology, this section illustrates how life evolves as a cohesive narrative of adaptation and innovation.

At its core, symbiogenesis reflects the principle of cellular homeostasis, where organisms strive to maintain internal balance in response to external environmental pressures. This dynamic balance is not static but an evolving state that integrates energy conservation and adaptation. According to Torday (2015), cellular homeostasis drives evolution by fostering innovation through environmental curiosity—a biological imperative to explore, adapt, and optimize energy flows. This process allows organisms to assimilate environmental challenges into their physiology, creating a resilient and adaptable biosystem.

The concept of symbiogenesis was first formalized by Sagan (1967), who identified it as a key driver of evolutionary complexity. For example, the integration of prokaryotic cells into early eukaryotic organisms to form mitochondria exemplifies how distinct entities can merge to enhance survival and functionality. This process not only resolved immediate environmental threats but also provided the energetic foundation for multicellular life, thereby weaving individual evolutionary events into a cohesive history.

Symbiogenesis also reframes the relationship between organisms and their environments. Unlike Darwinian selection, which prioritizes reproductive success, symbiogenesis highlights reciprocal environmental responsiveness. Gould and Vrba (1982) described this as an evolutionary exaptation—where historical traits are repurposed for new functions. For instance, structures like the lung alveolus originally evolved for gas exchange but later adapted to facilitate the demands of endothermy. These adaptations illustrate how symbiogenesis integrates environmental pressures into physiological systems, enhancing complexity and resilience.

Moreover, symbiogenesis bridges biology and cosmology by mirroring universal patterns of organization. Cellular homeostasis parallels the conservation laws of physics, such as energy conservation, which govern the cosmos. The ability of life to maintain order amidst entropy aligns with the fundamental thermodynamic principles that shape the universe. This connection situates symbiogenesis as a process that reflects not only the dynamics of life but also the broader logic of the cosmos.

Torday and Rehan (2017) propose that symbiogenesis is essential for understanding evolution as a holistic process. It offers a framework for viewing life as a continuum, where cellular and systemic adaptations are woven into the fabric of a larger cosmic narrative. This perspective emphasizes that life is not merely a product of random mutations and selection but a dynamic interplay of cooperation, innovation, and environmental integration.

In conclusion, symbiogenesis serves as the nexus of life and cosmos, uniting biological processes with universal laws. It challenges reductionist views of evolution, offering a holistic framework that highlights the interconnectedness of life's evolutionary history with the broader dynamics of the universe. This principle underscores the profound unity between organisms and their environments, positioning life as an integral part of the cosmic order.

## Phenotype as an Agent of Niche Construction

The phenotype, the observable traits and behaviors of an organism, plays a pivotal role in evolutionary innovation by actively shaping and responding to its environment. Unlike traditional views that consider the environment as a passive backdrop, the phenotype emerges as an active participant in niche construction, driving evolutionary dynamics. This section explores how phenotypic expression facilitates adaptation, influences environmental interactions, and integrates epigenetic data to sustain evolutionary trajectories.

The concept of niche construction highlights the agency of the phenotype in modifying environmental conditions to suit the organism's survival needs. Torday and Miller (2016) emphasize that phenotypic expression is not merely a reflection of genetic programming but a dynamic interface between the genome and environment. Organisms acquire epigenetic marks—heritable chemical modifications on DNA—in response to environmental stressors. These marks alter gene expression without changing the underlying DNA sequence, allowing the phenotype to adapt rapidly to changing conditions. This process exemplifies how phenotypes serve as mediators of evolutionary innovation, embedding environmental experiences into the evolutionary fabric.

An illustrative example of this dynamic is the obligate return of multicellular organisms to their unicellular state during reproduction. Matikainen et al. (2021) describe this process as an opportunity for epigenetic data retrieval, where unicellular gametes serve as repositories of environmental and genetic information. By reverting to a simplified cellular state, organisms can access and integrate accumulated evolutionary knowledge, ensuring that their offspring are primed for survival in the prevailing environmental context. This cyclical transition underscores the phenotype's role in preserving and transmitting adaptive strategies across generations.

The water-to-land transition provides a compelling case study of how phenotypic adaptations drive niche construction. During this critical evolutionary leap, vertebrates faced novel environmental challenges, such as increased gravitational forces and fluctuating oxygen availability. Gene duplications, such as those of the parathyroid hormone-related protein (PTHrP) receptor, enabled vertebrates to develop skeletal stiffening and lung alveolarization to adapt to terrestrial environments (Rubin et al., 2004). These mechanistic changes highlight the phenotype's capacity to actively reshape its environment by modifying its physiology to meet new demands.

The role of the phenotype in niche construction also extends to broader ecological impacts. By altering the environment, phenotypes create feedback loops that influence both individual survival and community dynamics. This bidirectional relationship between organism and environment demonstrates the interplay between biological adaptation and ecosystem evolution. For example, as early land vertebrates adapted to terrestrial life, they simultaneously transformed their habitats, creating conditions for further evolutionary diversification.

In conclusion, the phenotype is not a passive byproduct of genetic expression but an active agent of evolutionary change. Through niche construction, phenotypic adaptations drive the integration of environmental pressures into the evolutionary process, fostering innovation and resilience. By acquiring epigenetic marks, facilitating unicellular-to-multicellular transitions, and enabling mechanistic adaptations like those observed in the water-to-land transition, the phenotype exemplifies the dynamic interplay between organisms and their environments. This perspective underscores the centrality of phenotypic agency in shaping the trajectory of life's evolutionary narrative.

## The Cellular and Atomic Homology

The profound parallels between the atom and the cell provide a unique framework for understanding life's origins and evolution. Both represent fundamental units of their respective domains—atoms in the physical realm and cells in the biological realm. This homology bridges inanimate and animate systems, linking them through principles like negentropy, homeostasis, and evolutionary adaptation. By examining these parallels, we uncover how life integrates cosmic and cellular logic into a cohesive narrative.

Atoms are governed by deterministic principles, dictated by their structure and interactions as defined in the Periodic Table of Elements. Similarly, cells operate through physiological mechanisms that maintain homeostasis, ensuring stability amidst fluctuating environments (Torday & Rehan, 2012). This shared emphasis on order contrasts with the chaos of entropy, highlighting both systems' capacity to harness energy and sustain organized complexity. In the atom, this manifests through chemical bonds, while in the cell, it is reflected in metabolic processes that counter entropy by maintaining internal equilibrium.

The synchronic and diachronic dimensions of cellular evolution further parallel the periodic table's organization. Synchronic processes represent the immediate, real-time functions of the cell, such as energy metabolism and molecular signaling, which mirror the atomic interactions occurring in chemical reactions. Diachronic processes, on the other hand, refer to the historical adaptations that have accumulated over evolutionary time, analogous to the periodic table's diagonal relationships that signify trends in elemental properties (Burbidge et al., 1957). These dual dimensions illustrate how the dynamic interplay of real-time and historical processes underpins both atomic and cellular systems.

Stellar nucleosynthesis, the cosmic process by which elements are forged within stars, serves as the foundational source of the elements essential for life. The synthesis of hydrogen and helium during the Big Bang and subsequent stellar processes created heavier elements like carbon, oxygen, and nitrogen. These elements form the building blocks of cellular structures, such as lipids, proteins, and nucleic acids. Tegmark (2014) highlights how this cosmic lineage imbues biological systems with a logical structure derived from universal principles. For example, the carbon backbone's versatility allows for the complex molecular configurations necessary for cellular function, echoing the inherent adaptability of atomic structures.

The connection between atoms and cells extends beyond their structural and functional parallels to encompass their role in the broader cosmos. Just as atoms reflect the forces governing the universe, such as gravity and electromagnetism, cells embody the principles of adaptation and resilience that drive evolution. This shared narrative underscores the continuum between the inanimate and animate, positioning life as an emergent property of universal laws.

In conclusion, the homology between atoms and cells provides a compelling framework for understanding life's origins and evolution. By linking deterministic atomic principles with cellular homeostasis, and by paralleling synchronic and diachronic dimensions, this analogy bridges physical and biological realms. Moreover, the cosmic roots of cellular logic, as revealed through stellar nucleosynthesis, emphasize the deep interconnectedness of life and the universe. This perspective not only enriches our understanding of biology but also situates it within a broader cosmological context.

## Consciousness as the Aggregate of Evolutionary Adaptations

Consciousness emerges as an integrated product of evolutionary and cosmic processes, reflecting the challenges posed by the environment and the adaptive mechanisms that sustain life. It is a phenomenon shaped by the intricate interplay between physiology, cellular memory, and the universal principles governing existence. This section explores how consciousness embodies both local and nonlocal phenomena, seamlessly uniting the mechanistic and emergent aspects of life.

The foundation of consciousness lies in physiology's ability to adapt to existential challenges presented by the environment. Stephen Hawking (1998) emphasized the cosmos's finely tuned parameters, suggesting that the universe is inherently structured to foster complexity. In the context of life, this complexity is mirrored in the mechanisms of homeostasis, where cells and organisms continuously respond to external pressures while maintaining internal equilibrium. Cellular processes, such as energy metabolism and stress responses, serve as a microcosmic reflection of the universe's broader adaptive dynamics, illustrating how physiology integrates environmental inputs into coherent states of awareness.

Experimental evidence from microgravity studies provides compelling insights into the role of cellular memory in maintaining phenotypic identity. In the absence of gravitational forces, cells demonstrate a tendency to lose their specialized characteristics, reverting to more primitive states (Engle et al., 1980; Torday, 2003). This phenomenon underscores the critical role of environmental factors, such as gravity, in stabilizing cellular functions and preserving evolutionary memory. Cellular memory serves as a repository of past adaptations, enabling organisms to navigate present challenges and anticipate future needs. By maintaining a record of environmental interactions, cells contribute to the formation of a coherent consciousness that integrates synchronic (real-time) and diachronic (historical) dimensions.

Consciousness is not merely a product of local phenomena confined to the cellular or physiological level; it also encompasses nonlocal aspects that transcend the mechanistic framework of life. Roger Penrose (1996) and Kurt Gödel (1931) proposed that understanding complex systems requires acknowledging the limitations of formal logic and incorporating elements that exist beyond conventional boundaries. In biological terms, this suggests that consciousness arises from the integration of deterministic cellular processes with emergent properties that reflect the universe's nonlocal nature. For instance, the phenomenon of quantum entanglement in cellular signaling demonstrates how spatially separated components can remain interconnected, offering a model for understanding the holistic nature of consciousness.

The interplay between local and nonlocal phenomena in consciousness is further exemplified by the concept of cellular homeostasis, which operates both mechanistically and adaptively. At the local level, cells employ precise biochemical pathways to maintain balance, responding dynamically to stimuli. At the nonlocal level, these processes align with universal principles, such as entropy reduction and energy conservation, positioning life as a bridge between the deterministic laws of physics and the emergent complexities of biology.

In conclusion, consciousness represents the aggregate of evolutionary adaptations that integrate local physiological processes with nonlocal cosmic principles. By reflecting the environment's challenges through physiology, preserving evolutionary memory, and embodying emergent properties, consciousness transcends reductionist explanations. It is a unifying phenomenon that captures the essence of life's adaptive journey, linking the microcosm of cellular processes to the macrocosm of universal dynamics.

## Holism, Quantum Mechanics, and Cellular Integration

The synthesis of symbiogenesis, quantum mechanics, and cellular biology reveals life as a unified, holistic system transcending traditional disciplinary boundaries. This final section explores the interconnectedness of life, demonstrating how evolutionary processes, quantum phenomena, and cellular dynamics converge to form a cohesive understanding of consciousness and the cosmos.

The homology between symbiogenesis and quantum entanglement exemplifies life's fundamental unity. Symbiogenesis—the process by which cells incorporate environmental threats into their physiology—is analogous to quantum entanglement, where particles become interdependent regardless of spatial separation. Both mechanisms illustrate the integration of local and nonlocal dynamics, underscoring the interconnected nature of life (Torday, 2023). In this framework, cellular adaptation mirrors quantum processes, with each environmental challenge absorbed and transformed into a physiological advantage. Such a perspective transcends reductionist scientific models, positioning life as a system that integrates mechanistic and emergent phenomena.

Holism offers profound insights into consciousness, framing evolution and the periodic table as complementary topologies. Maturana and Varela's (1980) autopoiesis theory describes living systems as self-sustaining and self-creating entities, emphasizing their ability to maintain coherence through constant interaction with the environment. This concept aligns with the periodic table's synchronic (horizontal relationships) and diachronic (diagonal trends) dimensions, which mirror the synchronic and diachronic aspects of cellular evolution. Just as the elements are organized into predictable patterns, cellular processes reflect a similar topology of adaptive responses and evolutionary continuity.

The implications of holism extend beyond biology, offering a universal framework that bridges the unity of life with the cosmos. Smuts' (1926) philosophy of holism posits that the whole is greater than the sum of its parts—a concept that resonates with the integration of biological, physical, and quantum principles. Cellular processes such as homeostasis exemplify this integration, maintaining stability in the face of environmental flux. When viewed through a holistic lens, life becomes a dynamic system capable of uniting disparate scientific paradigms.

Quantum mechanics further enriches this understanding by illuminating the probabilistic and interconnected nature of existence. Cellular behavior, such as energy exchange and signaling, reflects quantum phenomena like wave-particle duality and entanglement. These processes enable cells to adapt to environmental changes while maintaining systemic coherence, providing a mechanistic foundation for consciousness. The holistic integration of these principles underscores the unity of life and the cosmos, suggesting that consciousness is an emergent property of this interplay.

This perspective offers profound implications for addressing fundamental questions in science, philosophy, and consciousness studies. By bridging biology, physics, and philosophy, holism provides a roadmap for exploring the unity of life, emphasizing the inseparability of its components. Evolution, when viewed through this lens, is not merely a process of adaptation but a continuous dialogue between life and the universe, culminating in the emergence of consciousness.

In conclusion, the synthesis of symbiogenesis, quantum mechanics, and holism reveals life as an integrated system that transcends traditional scientific frameworks. This understanding positions biology, physics, and philosophy as complementary disciplines, offering a unified approach to unraveling the mysteries of life and consciousness.

## References

Burbidge, E. M., Burbidge, G. R., Fowler, W. A., & Hoyle, F. (1957). Synthesis of the elements in stars. *Reviews of Modern Physics, 29*, 548–650.

Engle, M. J., Sanders, R. L., & Douglas, W. H. (1980). Type II alveolar cells in organotypic culture. *Biochimica et Biophysica Acta, 617*, 225–236.

Gödel, K. (1931). Über formal unentscheidbare Sätze der Principia Mathematica und verwandter Systeme I. *Monatshefte für Mathematik und Physik, 38*, 173–198.

Gould, S. J., & Vrba, E. S. (1982). Exaptation—A missing term in the science of form. *Paleobiology, 8*, 4–15.

Hawking, S. (1998). *A brief history of time*. Bantam Books.

Matikainen, N., Pekkarinen, T., & Ryhänen, E. M. (2021). Physiology of calcium homeostasis: An overview. *Endocrinology and Metabolism Clinics of North America, 50*, 575–590.

Maturana, H. R., & Varela, F. J. (1980). *Autopoiesis and cognition: The realization of the living*. Reidel Publishing Company.

Penrose, R. (1996). On gravity's role in quantum state reduction. *General Relativity and Gravitation, 28*, 581–600.

Rubin, L. P., Kovacs, C. S., & De Paepe, M. E. (2004). Arrested pulmonary alveolar cytodifferentiation and defective surfactant synthesis in mice missing the gene for parathyroid hormone-related protein. *Developmental Dynamics, 230*, 278–289.

Sagan, L. (1967). On the origin of mitosing cells. *Journal of Theoretical Biology, 14*, 255–274.

Smuts, J. C. (1926). *Holism and evolution*. Macmillan.

Tegmark, M. (2014). *Our mathematical universe: My quest for the ultimate nature of reality*. Knopf.

Torday, J. S. (2003). Parathyroid hormone-related protein is a gravisensor in lung and bone cell biology. *Advances in Space Research, 32*, 1569–1576.

Torday, J. S. (2015). Homeostasis as the mechanism of evolution. *Biology (Basel), 4*, 573–590.

Torday, J. S. (2023). Consciousness, embodied quantum entanglement. *Progress in Biophysics and Molecular Biology, 177*, 125–128.

Torday, J. S., & Miller, W. B. (2016). The resolution of ambiguity as the basis for life. *Progress in Biophysics and Molecular Biology, 122*(1), 1–8.

Torday, J. S., & Rehan, V. K., (2012). *Evolutionary biology, cell-cell communication and complex disease*. Wiley, Hoboken.

Torday, J. S., & Rehan, V. K. (2017). *Evolution, the logic of biology*. Wiley.

# Chapter 19
# Scientific Proof of the Sacred in Life

The quest to understand life's origins and its profound interconnectedness with the cosmos has long been the domain of both science and spirituality. In this chapter, we explore how modern scientific discoveries offer empirical evidence that aligns with ancient spiritual insights, revealing the sacredness embedded in the fabric of life. Far from being separate realms, science and spirituality emerge as complementary perspectives, united by the pursuit of understanding the essence of existence.

This chapter delves into the intrinsic order and purpose in nature, uncovering the sacred geometry of life through mathematical and biological frameworks. From the golden ratio evident in natural growth patterns to the fractal geometries of physiological structures, these patterns reflect a universal intelligence operating across scales. Such revelations challenge reductionist paradigms, suggesting that life's complexity and coherence are far more than random occurrences—they point to an underlying cosmic order.

A pivotal theme in this chapter is the concept of holism, which positions life as a unified system deeply embedded in its environment. Through the lens of symbiogenesis and evolutionary biology, we uncover how life evolves by integrating environmental challenges, forging symbiotic relationships that underscore the sacred interdependence of all living beings. These processes reflect a larger principle of reciprocity, in which life sustains and is sustained by its surroundings, mirroring spiritual notions of harmony and balance.

Quantum mechanics, another cornerstone of this exploration, bridges the gap between the physical and the metaphysical. By illuminating phenomena like quantum entanglement and the 20,164, it becomes clear that the universe operates not only through deterministic laws but also through dynamic, interconnected relationships. The role of consciousness in this framework becomes undeniable, offering a scientific basis for the sacred role of awareness in shaping reality.

Lastly, this chapter addresses the implications of these findings for our understanding of existence. By uniting the material and the spiritual, science provides a pathway to rediscovering the sacred within life itself. This perspective does not



J. S. Torday, R. G. Sacco, *Conscious Life*,
https://doi.org/10.1007/978-3-031-80500-4_19

merely enrich scientific inquiry; it transforms how we view ourselves, our environment, and our responsibilities as stewards of the planet. In connecting cellular evolution, quantum mechanics, and cosmology, this chapter reveals that life is not merely an accident of biology—it is a profound expression of universal principles that are, in essence, sacred.

Through this synthesis, we find that the sacred in life is not a matter of belief but a reality observable through the very mechanisms of nature. As we come to see science and spirituality as intertwined, a new understanding of existence emerges—one that celebrates the unity of all things and the divine intelligence that animates life.

## The Emergence of Holism and Its Scientific Basis

Holism, the principle that systems and their properties should be viewed as wholes rather than merely as collections of parts, serves as a profound bridge between reductionist science and the innate human quest for unity in life and the cosmos. This section explores the scientific underpinnings of holism, tracing its origins, experimental validations, and mechanistic explanations.

The term "holism" was first articulated by Jan Smuts in his seminal work Holism and Evolution (1926), where he proposed that the evolution of the universe exhibits a tendency toward the formation of wholes, from atoms to living organisms to societies. Smuts drew philosophical inspiration from Plotinus's concept of "The One," a metaphysical idea that all of existence emanates from a singular, indivisible source. In this framework, holism transcends mere philosophy, offering a scientific lens through which to view the interconnectedness of life and its environment. This perspective rejects reductionism, which isolates components without recognizing their dynamic interactions, and instead emphasizes systems as integrated and indivisible entities.

Experimental evidence for holism finds a compelling foundation in microgravity studies. Gravity, an inanimate force, profoundly influences cellular differentiation, illustrating the continuum between the physical and biological realms. Research conducted during microgravity experiments has shown that cells lose their phenotypic identity in the absence of gravitational forces, reverting to a more primitive state (Torday, 2003). This finding underscores gravity's role in shaping the emergent complexity of life, linking it to evolutionary processes that are both synchronic (moment-to-moment adaptations) and diachronic (historical transformations over evolutionary time). In this sense, holism emerges as an observable phenomenon: even in the simplest cellular systems, the whole is far greater than the sum of its parts.

The mechanistic basis of holism is further elaborated through the concepts of serial homeostasis and optimal entropy. Serial homeostasis refers to the sequential regulation of physiological systems to maintain internal stability, a principle that scales up from individual cells to entire ecosystems (Torday, 2024c). For example, the adaptive changes required for the water-to-land transition in vertebrate evolution illustrate how homeostasis drives integrative complexity. This process is intrinsically tied to entropy, the measure of disorder in a system. Holism suggests that life exists in a state of "optimal entropy," where systems balance order and chaos to sustain themselves. This balance is not random but emerges from the iterative interactions of components working together as a whole.

Holism thus bridges the animate and inanimate worlds, providing a framework that unites the reductionist understanding of biological mechanisms with the integrative, systemic view of life as a cohesive entity. By recognizing the interdependence of physical forces, cellular processes, and ecological systems, holism offers a scientific basis for understanding the unity of life and its place within the cosmos. As both a philosophical and empirical principle, holism reshapes our understanding of existence, emphasizing the interconnectedness that defines all levels of reality.

## Symbiogenesis as a Driver of Holistic Evolution

Symbiogenesis, the process by which separate organisms merge to form a new, more complex entity, is not only a biological phenomenon but also a profound mechanism for understanding the interconnectedness of life and its relationship to cosmic processes. This section explores how symbiogenesis integrates environmental challenges into life's physiology, linking cellular evolution to universal principles of organization.

Proposed by Lynn Margulis and earlier conceptualized by Konstantin Mereschkowski, symbiogenesis serves as a cornerstone for understanding evolutionary complexity. Carl Sagan (1967) further framed this process as central to the development of life, highlighting its capacity to integrate disparate elements into cohesive physiological systems. By merging previously independent life forms, such as mitochondria and early eukaryotic cells, symbiogenesis facilitated the emergence of multicellular organisms. These mergers represent not just the addition of capabilities but a holistic transformation that allowed life to adapt to increasingly complex environments. In this sense, symbiogenesis embodies the evolutionary principle that survival is achieved not by competition alone but by cooperation and integration.

The mechanisms of symbiogenesis mirror those observed in quantum entanglement, where particles maintain a connected state regardless of distance, acting as if they are part of a unified whole. Torday (2023) posits that symbiogenesis and quantum entanglement are homologous processes, both of which underpin life's ability to maintain homeostasis in a dynamic and unpredictable cosmos. Just as entangled particles respond instantaneously to changes in their paired counterparts, symbiogenesis allows life to incorporate environmental pressures into its physiology, preserving balance and promoting adaptation. This parallel suggests that life's evolution is not merely biological but fundamentally quantum in nature, with mechanisms like symbiogenesis bridging the local and nonlocal dimensions of existence.

A striking example of symbiogenesis as a cosmic process is the assimilation of environmental elements that align with the logical structure of the periodic table. Hoyle (1946) emphasized that the elements essential for life—carbon, hydrogen, oxygen, and nitrogen—were forged in the interiors of stars through stellar nucleosynthesis. Symbiogenesis incorporates these elements into cellular structures and processes, effectively inheriting the "cosmic logic" embedded in the periodic table. For example, the integration of iron into hemoglobin molecules for oxygen transport reflects not only an evolutionary adaptation but also the utilization of elemental properties shaped by the universe's physical laws.

Through symbiogenesis, life becomes a reflection of the cosmos, inheriting its order and adapting it to the ever-changing demands of Earth's environment. This perspective frames life not as an isolated phenomenon but as a continuum of the universe's creative processes, where cellular homeostasis mirrors the balance of forces governing cosmic systems. As a mechanism for uniting the animate and inanimate, symbiogenesis exemplifies holism, offering insights into how life transcends reductionist frameworks to achieve integrative complexity.

## Synchronic and Diachronic Processes in Evolution

The processes of evolution unfold across two interwoven dimensions: synchronic (real-time adaptations) and diachronic (long-term changes over evolutionary history). Understanding these dimensions reveals how life's complexity emerges through the interplay of cellular, organismal, and cosmic dynamics. This section frames evolution as a continuum, with cellular communication serving as the fundamental bridge between these temporal scales.

Ontogeny (the development of an organism) and phylogeny (the evolutionary history of a species) exemplify synchronic and diachronic processes, respectively. Cellular communication plays a pivotal role in aligning these dimensions, ensuring that immediate adaptations integrate seamlessly with long-term evolutionary trajectories. Torday and Rehan (2012) emphasize that the same cellular mechanisms governing development, such as intercellular signaling and homeostasis, are repurposed over evolutionary time to drive species-wide changes. For instance, the alveolar structure of the lung reflects both the immediate need for efficient gas exchange and the diachronic evolutionary response to oxygen fluctuations in Earth's atmosphere.

The synchronic and diachronic interplay in evolution finds mathematical expression in the Fibonacci sequence and fractal geometry. These patterns exemplify emergence, where simple iterative processes give rise to complex, self-similar structures across scales. Blaisdell et al. (2013) illustrate how fractal physiology governs systems like the branching of blood vessels and neural networks, optimizing energy distribution and communication. Similarly, Sacco and Torday (2023) propose that the Fibonacci sequence represents the temporal unfolding of evolutionary processes, linking past and future adaptations. These mathematical models demonstrate how life mirrors cosmic patterns, embedding universal principles within biological systems.

Mendeleev's periodic table offers a compelling analogy for understanding these processes. Just as the periodic table organizes elements by their reactivity and relationships, evolution organizes life by its adaptability and interdependence. The reactivity of elements, such as the affinity of carbon to form diverse molecular bonds, parallels the adaptability of cellular systems in response to environmental changes. This analogy underscores the synchronic aspect of immediate reactivity while capturing the diachronic emergence of complexity through the accumulation of molecular and physiological innovations.

The integration of synchronic and diachronic processes reveals evolution as a dynamic system, where present adaptations feed into future trajectories and past innovations inform real-time survival. Cellular communication serves as the mechanism that ensures coherence across these temporal scales, linking the immediate needs of an organism to the evolutionary history of its lineage. This dual perspective provides a holistic framework for understanding the unity of life, bridging the microcosm of cellular processes with the macrocosm of cosmic patterns.

By framing evolution as the interplay of synchronic and diachronic dimensions, we gain deeper insights into the mechanisms that sustain life and consciousness. This perspective emphasizes the continuity of biological and cosmic processes, highlighting the fundamental interconnectedness of all systems.

## Holism as a Framework for Spirituality and Morality

Holism, the recognition of interconnectedness and unity in life, provides a scientific and philosophical foundation for understanding spirituality and morality. By emphasizing the interdependence of all living systems, holism reframes our relationship with the cosmos, fostering empathy, collaboration, and ethical behavior. This perspective bridges the divide between scientific inquiry and spiritual understanding, offering a coherent framework for navigating life's complexities.

Holism as a guiding principle recognizes that life operates not as a collection of isolated components but as a dynamic, interconnected system. Krishnamurti and Bohm (1983) argue that this perspective nurtures empathy and ethical behavior, as it reveals the inherent unity among individuals, societies, and ecosystems. Understanding the mutual dependence of all life forms encourages collaboration over competition, framing morality as an extension of biological and cosmic processes. For example, cellular symbiosis, where organisms integrate environmental challenges into their physiology, mirrors ethical collaboration in human societies, underscoring the shared principles of survival and coexistence.

Spiritual awareness emerges naturally from this holistic understanding. Holism posits that individual consciousness is inseparable from the larger web of existence, connecting humans to the cosmos through shared physiological and environmental processes. Kauffman (2008) describes this connection as "reinventing the sacred," where the complexity and order inherent in life inspire a sense of awe traditionally associated with spirituality. By acknowledging life's intricate interdependence, holism transcends reductionist interpretations of spirituality, grounding it in observable phenomena such as cellular communication, quantum entanglement, and energy regulation.

Scientific holism also provides a robust framework for interpreting morality, reframing ethical behavior as a natural consequence of life's interconnectedness. The interdependence observed in ecosystems, from nutrient cycling to symbiotic relationships, reflects a universal principle of mutual support and reciprocity. Torday (2023) extends this concept to human behavior, proposing that moral actions arise from our biological imperative to maintain homeostasis and balance within the larger system. This perspective situates morality within the continuum of evolutionary processes, emphasizing that ethical behavior sustains not only individuals but also the collective integrity of life.

Furthermore, holism challenges reductionist interpretations of life that focus solely on material explanations, offering instead a vision of life as an interconnected system sustained by symbiotic and quantum processes. By acknowledging the role of phenomena such as quantum entanglement and nonlocal consciousness, holism expands the boundaries of traditional science, integrating spiritual and moral dimensions into its explanatory framework. This holistic approach reframes questions of purpose and meaning, aligning them with the principles of unity and interdependence observed across all scales of existence.

Holism thus serves as a bridge between science, spirituality, and morality, offering a unified framework for understanding life's interconnected nature. By recognizing the unity inherent in all systems, it inspires empathy, collaboration, and ethical behavior, providing a pathway for addressing personal, societal, and ecological challenges.

## Implications for Science, Society, and the Cosmos

Holism, as a scientifically grounded perspective, offers profound implications for understanding life's purpose, addressing societal challenges, and bridging the gap between science and spirituality. By framing life as a product of cosmic singularity with a nondual origin, holism provides a unifying framework that connects individual existence with the broader cosmos.

The philosophical implications of holism begin with its acknowledgment of life as emerging from a singularity—both a physical and conceptual origin point. Tegmark (2014) argues that the universe is fundamentally mathematical, where life is an expression of the underlying logic and unity of the cosmos. This nondual perspective suggests that distinctions between animate and inanimate, conscious and unconscious, or material and spiritual are constructs of human perception rather than inherent divisions. Torday (2019) extends this idea by linking life's emergence to the interplay of energy flows, entropy regulation, and evolutionary adaptation, framing life as an integral part of the cosmos rather than a separate phenomenon.

Recognizing our interconnectedness with the cosmos has transformative implications for addressing societal and environmental challenges. Erickson (2007) proposes that the realization of life's unity fosters a sense of collective responsibility, emphasizing the need for sustainable practices and cooperative approaches to global issues. For example, the concept of symbiogenesis illustrates how life evolves through integration and mutualism, providing a model for societal collaboration and problem-solving. By adopting a holistic view, society can move beyond competitive paradigms and embrace solutions that balance individual needs with collective well-being.

The alignment of science with spirituality through holism bridges historically disparate fields, offering a comprehensive perspective that acknowledges both empirical evidence and existential meaning. Holism does not merely unify scientific disciplines but also reconciles scientific inquiry with humanity's deeper quest for purpose. The recognition of life's interdependence fosters a spiritual awareness that transcends religious boundaries, emphasizing shared existence and universal principles. Torday (2019) suggests that this alignment can lead to a deeper understanding of morality, ethics, and the role of humanity within the larger ecological and cosmic framework.

Holism also reframes life's purpose by situating it within the continuum of cosmic evolution. The interconnectedness of all systems—biological, ecological, and cosmic—reveals that life is not a random occurrence but an emergent property of the universe's fundamental laws. This perspective aligns with Tegmark's (2014) proposition that consciousness and existence are natural extensions of the cosmos's mathematical structure, offering a meaningful narrative for humanity's place in the universe.

In conclusion, holism serves as a scientifically and philosophically grounded perspective for understanding life's unity and purpose. By recognizing the interconnectedness of all systems, it provides a framework for addressing societal and environmental challenges, aligning science with spirituality, and offering humanity a deeper sense of belonging within the cosmos. This unified perspective not only enhances our understanding of life but also inspires a collective effort to create a sustainable and harmonious future.

## References

Blaisdell, A. P., Pottenger, B. C., & Torday, J. S. (2013). From heart beats to health recipes: The role of fractal physiology in the Ancestral Health movement. *Journal of Evolution and Health, 1*(2).

Erickson, H. L. (2007). Philosophy and theory of holism. *Nursing Clinics of North America, 42*, 139–163.

Hoyle, F. (1946). The synthesis of the elements from hydrogen. *Monthly Notices of the Royal Astronomical Society, 106*, 343–383.

Kauffman, S. A. (2008). *Reinventing the sacred: A new view of science, reason, and religion*. Basic Books.

Krishnamurti, J., & Bohm, D. (1983). *Discussions on consciousness and the cosmos*. Brockwood Park Dialogues.

Sacco, R. G., & Torday, J. S. (2023). Systems biology of human aging: A Fibonacci time series model. *Progress in Biophysics and Molecular Biology, 177*, 24–33.

Sagan, L. (1967). On the origin of mitosing cells. *Journal of Theoretical Biology, 14*, 255–274.

Smuts, J. C. (1926). *Holism and evolution*. Macmillan.

Tegmark, M. (2014). *Our mathematical universe*. Vintage.

Torday, J. S. (2003). Parathyroid hormone-related protein is a gravisensor in lung and bone cell biology. *Advances in Space Research, 32*, 1569–1576.

Torday, J. S. (2019). The singularity of nature. *Progress in Biophysics and Molecular Biology, 142*, 23–31.

Torday, J. S. (2023). Consciousness, embodied quantum entanglement. *Progress in Biophysics and Molecular Biology, 177*, 125–128.

Torday, J. S. (2024c). The synchronic, diachronic cell as the holism of consciousness. *Progress in Biophysics and Molecular Biology, 188*, 19–23.

Torday, J. S., & Rehan, V. K. (2012). *Evolutionary biology, cell-cell communication, and complex disease*. Wiley.

# Chapter 20
# The Cosmos as a Map for Consciousness

The cosmos has long captivated human curiosity, serving as both a source of wonder and a mirror for introspection. In this chapter, we explore the profound relationship between the universe's structure and the emergence of consciousness, framing the cosmos as a living map that reflects and informs the nature of conscious awareness. From the intricate patterns of stellar nucleosynthesis to the recursive symmetries of cellular processes, the cosmos offers a template for understanding the interconnectedness of all existence.

Life, as a manifestation of cosmic principles, is not a random occurrence but a continuation of the universe's logic. This chapter delves into how the dynamic interplay of energy, matter, and information forms a continuum that bridges the inanimate and animate, ultimately giving rise to consciousness. Drawing on insights from quantum mechanics, cellular biology, and cosmology, we frame the cosmos as both a physical reality and a metaphorical framework that maps the trajectory of consciousness from its origins to its potential future.

The chapter begins by examining the cosmic origins of life, highlighting the role of stellar processes in creating the elements essential for cellular structures. It then explores how the same principles that govern the cosmos—such as symmetry, entropy regulation, and feedback loops—are mirrored in biological systems, providing the foundation for emergent phenomena like consciousness. Through this lens, we consider the universe not as a distant backdrop but as an active participant in the evolution of life and awareness.

We also address the philosophical implications of viewing the cosmos as a map for consciousness. How does understanding the universe's interconnected design shape our sense of identity and purpose? What can the cosmos teach us about the unity of existence and the nature of perception? By linking the material and metaphysical dimensions, this chapter provides a holistic perspective on the role of consciousness as both a product and a reflection of the universe.

Finally, this chapter emphasizes the transformative potential of aligning our understanding of consciousness with the cosmos. By recognizing the shared



J. S. Torday, R. G. Sacco, *Conscious Life*,
https://doi.org/10.1007/978-3-031-80500-4_20

principles that govern life and the universe, we can cultivate a deeper appreciation for our place within this vast interconnected system. The cosmos, in this view, becomes not only a map but also a guide, offering insights into the nature of reality and the pathways to self-awareness and collective evolution.

Through scientific exploration and philosophical reflection, *The Cosmos as a Map for Consciousness* invites us to see the universe as a living blueprint, one that charts the profound unity between life, mind, and the cosmos itself.

## Mapping Evolution onto Cosmic Principles

The relationship between evolution and cosmic phenomena offers profound insights into the unity of life and the universe. Evolution, traditionally seen as a biological process, can be reframed as a continuum that mirrors the fundamental principles governing the cosmos. By examining how life's progression parallels the structures and dynamics of the universe, this section lays the groundwork for understanding evolution as a reflection of cosmic logic.

### *Evolution and the Synchronic-Diachronic Framework*

Evolution can be mapped onto the periodic table's synchronic (real-time) and diachronic (spatial-temporal) dimensions, which provide a model for understanding the interplay between stability and change. The synchronic aspect reflects real-time processes, such as cellular communication and homeostasis, while the diachronic dimension captures evolutionary history through the accumulation of adaptations over time. These principles echo the dual nature of cosmic phenomena: the synchronic stability of celestial structures, such as stars and galaxies, and the diachronic expansion of the universe itself (Torday, 2022; Burbidge et al., 1957).

The periodic table, as a systematic arrangement of elements, mirrors the iterative processes of cellular evolution. Just as elements are organized based on atomic structure and reactivity, evolutionary adaptations follow patterns of functional integration and environmental responsiveness. This homology between the periodic table and biological evolution underscores the universality of organizational principles, linking the microscopic realm of cells to the macroscopic dynamics of the cosmos.

### *Micelles and the Space-Time Continuum*

The origins of life in micelles further illustrate the connection between evolution and cosmological principles. Micelles, spherical structures formed by lipid molecules in aqueous environments, represent the first step in cellular evolution. These structures exhibit remarkable homology with Einstein's theory of space-time, in which curvature and boundary conditions dictate interactions within the universe (Einstein, 1920; Torday & Rehan, 2012). Like space-time, micelles create a defined boundary that separates internal and external environments, enabling energy flows and the emergence of complexity.

The formation of micelles reflects the universe's inherent tendency toward order and organization despite entropy. By creating semipermeable boundaries, micelles laid the groundwork for cellular processes such as homeostasis and energy regulation, which are fundamental to life. This interplay between internal stability and external responsiveness mirrors the dynamics of the cosmos, where gravitational forces and energy flows sustain the structure of galaxies and stars.

## *Symbiogenesis: Bridging Biology and Cosmology*

Symbiogenesis serves as a conceptual bridge between evolutionary biology and cosmic phenomena. This process, which integrates environmental challenges into physiological systems, parallels the synthesis of elements during stellar nucleosynthesis (Sagan, 1967; Slijepčević, 2021). Just as stars produce heavier elements through nuclear fusion, symbiogenesis enables life to assimilate external stresses, transforming them into adaptive innovations.

The parallels between stellar and cellular processes highlight the shared logic that underpins evolution and cosmology. By viewing evolution through the lens of symbiogenesis, we can see life as an extension of cosmic processes, where the integration of matter and energy drives complexity and consciousness.

In framing evolution as a reflection of cosmic principles, this section reveals the profound interconnectedness between life and the universe, offering a unified perspective that transcends traditional disciplinary boundaries.

## Consciousness as the Integration of Cellular and Cosmic Signals

Consciousness, the enigmatic phenomenon that defines human experience, can be reimagined as an emergent property of the profound integration between cellular and cosmic processes. By examining its physiological, philosophical, and cosmological dimensions, this section offers a unifying framework for understanding consciousness as both a local and nonlocal phenomenon.

## *Cellular Foundations of Consciousness*

The physiological basis of consciousness is rooted in cell–cell communication, which emerged as a survival mechanism to maintain homeostasis and environmental adaptability. This communication, governed by biochemical signals and feedback loops, underpins the ability of organisms to perceive and respond to their surroundings. Torday (2020) suggests that this process originates from the Big Bang, conceptualized as the "endogenization" of cosmic energy into biological systems. The formation of lipid micelles, the precursors to cellular life, reflects this foundational integration, as their semipermeable boundaries encapsulate and regulate energy flows in a manner homologous to the cosmos itself.

Cellular communication creates a unified system capable of integrating vast amounts of information, laying the groundwork for the emergence of consciousness. Through the intricate interplay of synaptic activity, hormonal signaling, and metabolic regulation, the body generates a cohesive sense of self, bridging the microscopic realm of cells with the macroscopic dynamics of the universe.

## *The "Hard Problem" and Cosmic Integration*

David Chalmers (1995) famously articulated the "hard problem of consciousness," highlighting the difficulty of explaining how subjective experience (qualia) arises from physical processes. This puzzle can be approached by framing qualia as the integration of cellular and cosmic signals. Consciousness emerges from the alignment of cellular activities with universal laws, where the synchronic (moment-to-moment) and diachronic (evolutionary) dimensions of existence converge.

The physiological processes that generate qualia can be seen as reflections of cosmic phenomena. For instance, the rhythmic oscillations of neural activity mirror the cyclical patterns of the cosmos, such as planetary orbits and stellar lifecycles. Kuhn (2024) extends this perspective, proposing that consciousness acts as a bridge between local physiological processes and nonlocal cosmological principles, creating a unified experience of reality.

## *Bohm's Implicate and Explicate Orders*

The dual nature of consciousness as both local and nonlocal aligns with David Bohm's framework of the Implicate and Explicate Orders (Bohm, 1980). The Implicate Order represents the interconnected, nonlocal reality from which all phenomena arise, while the Explicate Order manifests as the observable, localized world. Consciousness embodies the dynamic interplay between these realms, as cellular processes translate cosmic signals into tangible experiences.

The synchronic aspect of consciousness, rooted in real-time cellular interactions, corresponds to the Explicate Order, where biological processes unfold visibly and measurably. In contrast, the diachronic dimension, encompassing evolutionary memory and nonlocal phenomena, aligns with the Implicate Order, reflecting the timeless, universal nature of consciousness.

Consciousness, when reframed as the integration of cellular and cosmic signals, transcends traditional disciplinary boundaries. By linking physiological processes to cosmological principles, this perspective provides a holistic understanding of consciousness as an emergent property of life and the universe. It challenges reductionist paradigms and opens pathways for interdisciplinary research, uniting biology, philosophy, and cosmology in the quest to unravel the mysteries of existence.

## Preventive Medicine and the Evolutionary Continuum

Preventive medicine, rooted in evolutionary biology, offers profound insights into understanding health and disease through the lens of evolutionary adaptations. By applying evolutionary developmental biology (evo-devo) principles, this section highlights the connection between evolutionary processes and medical innovation, emphasizing the continuum of ontogeny (developmental processes) and phylogeny (evolutionary history) as a framework for preventive healthcare.

### *Evo-Devo Principles and Cellular Regulatory Networks*

Evolutionary developmental biology (evo-devo) illuminates how species evolve through changes in developmental pathways, offering a robust model for understanding maladaptive traits. Cellular regulatory networks, which govern the formation and function of organs, are central to this process. These networks serve as both the architects and custodians of physiological systems, ensuring homeostasis while responding to environmental stresses.

Chronic lung diseases, for instance, can be traced back to disruptions in these regulatory networks. Torday and Rehan (2009a) argue that chronic pulmonary conditions, such as asthma and chronic obstructive pulmonary disease (COPD), are modern manifestations of evolutionary trade-offs. Cellular adaptations, such as those in the lung's alveolar structure, were optimized for specific environmental conditions. When those conditions shift, as in industrialized societies with high pollution levels, these adaptations may become maladaptive, leading to disease.

### *Vertical Integration and Predictive Modeling*

Preventive medicine benefits significantly from the vertical integration of evolutionary principles, linking molecular, cellular, and systemic levels of biology to understand disease mechanisms. This integration allows for the creation of predictive models that can anticipate disease onset based on evolutionary patterns. For example, Laland et al. (2008) highlight how understanding the evolutionary trajectory of genes and phenotypes provides predictive power in identifying populations at risk for certain conditions.

Torday (2018) extends this idea by proposing that preventive medicine should focus on understanding how evolutionary pressures shaped human physiology. Diseases often arise when there is a mismatch between our evolutionary heritage and modern environments. By identifying these mismatches, physicians can design interventions to prevent disease before it manifests. For instance, examining the evolutionary history of dietary adaptations could inform strategies for preventing metabolic disorders such as diabetes.

### *The Continuum Between Ontogeny and Phylogeny*

The relationship between ontogeny and phylogeny underscores the evolutionary continuum that shapes physiology. Ontogeny, the process of individual development, recapitulates key aspects of phylogeny, the evolutionary history of a species. This connection provides a deep understanding of how physiological systems have evolved to meet environmental demands.

For example, the lung's evolution from a swim bladder in fish to a highly efficient respiratory organ in mammals demonstrates this continuum. The study of these transitions informs medical approaches to respiratory diseases by highlighting the developmental processes that can be targeted for therapeutic interventions.

By leveraging the evolutionary continuum between ontogeny and phylogeny, preventive medicine can transform healthcare from reactive to proactive. Evo-devo principles offer a framework for understanding maladaptive traits, while vertical integration allows for the development of predictive models that anticipate disease. This approach bridges the gap between evolutionary biology and clinical practice, emphasizing that understanding our evolutionary past is essential for shaping a healthier future.

## Nonlocal Consciousness and Cosmic Laws

The concept of nonlocal consciousness challenges conventional understandings of life and its integration with the cosmos, emphasizing that cellular processes are deeply intertwined with universal principles. This section examines how cells, as the origins of consciousness, connect to cosmic laws through their physiological processes and explores the implications for understanding epigenetic inheritance, evolutionary adaptation, and human consciousness.

## *The Cell as the Origin of Consciousness*

Cells are not only the building blocks of life but also the fundamental originators of consciousness. Reber et al. (2024) argue that consciousness, often attributed to higher brain functions, emerges from cellular homeostasis and communication. Cellular processes such as signaling and feedback loops reflect a capacity for integrating local and nonlocal inputs, positioning the cell as a bridge between the microcosm of life and the macrocosm of the universe.

Physiologically, cells maintain homeostasis by responding to internal and external stimuli, a process that mirrors cosmic principles of equilibrium. This connection positions the cell as a microcosmic representation of universal laws, capable of navigating complexity through adaptive strategies.

## *Epigenetic Inheritance and Evolutionary Adaptations*

Epigenetic inheritance provides a compelling example of how cells integrate environmental influences into physiological and evolutionary frameworks. Through mechanisms such as DNA methylation and histone modification, cells encode environmental experiences into genetic material, enabling organisms to adapt across generations (Torday & Miller, 2018). These processes suggest a nonlocal dimension to cellular memory, where the influence of external factors persists beyond the immediate context.

For instance, environmental stressors such as temperature changes or nutrient scarcity can trigger epigenetic modifications that are passed to offspring, shaping their development and survival strategies. This ability to integrate environmental inputs aligns cellular behavior with broader cosmic principles, reinforcing the unity of life and the environment.

## *Human Consciousness and the Laws of Nature*

Human consciousness, as an emergent property of cellular processes, reflects the integration of environmental and cosmological influences. De Duve (1969) proposed that the principles governing the universe, such as thermodynamics and entropy, are mirrored in the physiological processes that sustain life. Human consciousness, in this context, can be seen as an advanced expression of the same universal laws that govern cellular homeostasis.

For example, the human brain, with its vast network of neurons and synapses, operates as a system of dynamic equilibrium. This balance allows for the integration of sensory inputs, cognitive functions, and emotional states, creating a unified experience of reality. Similarly, cellular homeostasis maintains physiological balance through feedback loops, illustrating a parallel between individual consciousness and universal processes.

Nonlocal consciousness, as rooted in cellular processes, provides a framework for understanding life as an interconnected system governed by cosmic laws. By integrating environmental and cosmological inputs through homeostasis, cells demonstrate a capacity for adaptation and memory that extends to human consciousness. This perspective not only redefines our understanding of life and evolution but also situates humanity within a continuum that spans from the cellular to the cosmic. Recognizing this unity offers profound implications for science, philosophy, and our relationship with the universe.

## Toward a Holistic Understanding of Life and the Cosmos

The quest for a holistic understanding of life and consciousness begins at the cellular level and extends to the vastness of the cosmos. This final section synthesizes the chapter's insights, emphasizing the interconnectedness of biological and cosmic cycles, the role of gravity in evolution, and the philosophical implications of framing life as a continuum within the universe.

### *The Zygote-to-Zygote Life Cycle as a Reflection of Cosmic Cycles*

Life's cyclical nature is exemplified by the zygote-to-zygote continuum, which mirrors cosmic cycles of creation and renewal. Torday and Rehan (2009b) frame this cycle as an energy flow system, prioritizing dynamic processes over static materialism. The zygote, as the origin of multicellular organisms, carries the evolutionary history encoded in cellular memory and epigenetic markers. This perspective shifts focus from individual components to the integrative processes that sustain life.

The parallels between biological cycles and cosmic phenomena are striking. Just as the universe expands and contracts through phases of energy transformation, life progresses through birth, growth, reproduction, and death, maintaining energy flows across generations. This cyclicality underscores the universality of energy as the foundational currency of existence, linking life to its cosmological origins.

## *Gravity as the Unifying Force in Cellular Evolution*

Gravity, a fundamental force in the cosmos, has played a pivotal role in shaping cellular evolution. Camirand et al. (2016) and Purevdorj-Gage et al. (2006) demonstrated that gravitational forces stabilize phenotypic expression, influencing differentiation and homeostasis. These effects are particularly evident in studies conducted in microgravity environments, where cells revert to more primitive states, highlighting gravity's role in maintaining the integrity of complex life forms.

From the unicellular state to the evolution of multicellular organisms, gravity has guided key adaptations, such as skeletal development and the emergence of bipedalism. This continuity illustrates how physical forces act as a unifying thread, connecting the microcosmic world of cellular processes to the macrocosmic dynamics of the universe.

## *Philosophical Implications: Life as a Continuum of the Cosmos*

The relationship between consciousness, evolution, and the cosmos invites profound philosophical reflection. Schrödinger (1944) described life as a system that maintains order against entropy, a concept that resonates with the idea of the universe as a self-organizing entity. Torday (2024) expands on this notion, suggesting that consciousness emerges as an extension of the universe's intrinsic drive for complexity and integration.

This perspective reframes humanity's place in the cosmos, emphasizing interconnectedness rather than separation. By viewing life as a continuum that bridges cellular processes and cosmic phenomena, we gain a deeper understanding of consciousness not as a localized phenomenon but as a universal principle.

Toward a holistic understanding of life and the cosmos, the zygote-to-zygote cycle, gravity's role in evolution, and the continuum of energy flows offer a unified framework that transcends disciplinary boundaries. Life is not an isolated occurrence but a dynamic interplay of forces and processes that reflect the universe's intrinsic logic. Recognizing this unity fosters a greater appreciation of our place within the cosmos and invites interdisciplinary approaches to unravel the mysteries of existence.

# References

Bohm, D. (1980). *Wholeness and the implicate order*. Routledge.

Burbidge, E. M., Burbidge, G. R., Fowler, W. A., & Hoyle, F. (1957). Synthesis of the elements in stars. *Reviews of Modern Physics, 29*, 547–650.

Camirand, A., et al. (2016). The role of parathyroid hormone-related protein in osteoblast response to microgravity. *PLoS One, 11*, e0160034.

Chalmers, D. J. (1995). Facing up to the problem of consciousness. *Journal of Consciousness Studies, 2*, 200–219.
De Duve, C. (1969). Evolution of the peroxisome. *Annals of the New York Academy of Sciences, 168*, 369–381.
Einstein, A. (1920). *Relativity: The special and general theory*. H. Holt and Company.
Kuhn, R. L. (2024). A landscape of consciousness. *Progress in Biophysics and Molecular Biology, 190*, 28–169.
Laland, K. N., et al. (2008). Evo-devo and niche construction. *Journal of Experimental Zoology B, 310*, 549–566.
Purevdorj-Gage, B., et al. (2006). Effects of low-shear modeled microgravity on yeast. *Applied and Environmental Microbiology, 72*, 4569–4575.
Reber, A. S., et al. (2024). The sensual cell. *Biosystems, 238*, 105197.
Sagan, L. (1967). On the origin of mitosing cells. *Journal of Theoretical Biology, 14*, 255–274.
Schrödinger, E. (1944). *What is life?* Macmillan.
Slijepčević, P. (2021). Serial endosymbiosis theory. *Biosystems, 202*, 104353.
Torday, J. S. (2018). Quantum mechanics predicts evolutionary biology. *Progress in Biophysics and Molecular Biology, 135*, 11–15.
Torday, J. S. (2020). Consciousness, Redux. *Medical Hypotheses, 140*, 109674.
Torday, J. S. (2022). Evolution, gravity, and the topology of consciousness. *Progress in Biophysics and Molecular Biology, 174*, 50–54.
Torday, J. S. (2024). The quantum cell. *Progress in Biophysics and Molecular Biology, 188*, 24–30.
Torday, J. S., & Miller, W. B. (2018). A systems approach to physiologic evolution. *Journal of Cellular Physiology, 233*, 162–167.
Torday, J. S., & Rehan, V. K. (2009a). Exploiting cellular-developmental evolution as the scientific basis for preventive medicine. *Medical Hypotheses, 72*, 596–602.
Torday, J. S., & Rehan, V. K. (2009b). Lung evolution as a cipher for physiology. *Physiological Genomics, 38*, 1–6.
Torday, J. S., & Rehan, V. K. (2012). *Evolutionary biology, cell-cell communication, and complex disease*. Wiley.

# Index




J. S. Torday, R. G. Sacco, *Conscious Life*,
https://doi.org/10.1007/978-3-031-80500-4

The manufacturer's authorised representative in the EU is Springer Nature Customer Service Centre GmbH, Europaplatz 3, 69115 Heidelberg, Germany. If you have any concerns regarding our products, please contact ProductSafety@springernature.com

Printed and bound by CPI Group (UK) Ltd, Croydon, CR0 4YY
01/10/2026
02225865-0007